Spatial Thinking in Environmental Contexts

Maps, Archives, and Timelines

Spatial Thinking in Environmental Contexts

Maps, Archives, and Timelines

Edited by
Sandra L. Arlinghaus, Joseph J. Kerski,
Ann Evans Larimore, Matthew Naud

CRC Press
Taylor & Francis Group
Boca Raton London New York

CRC Press is an imprint of the
Taylor & Francis Group, an **informa** business

CRC Press
Taylor & Francis Group
6000 Broken Sound Parkway NW, Suite 300
Boca Raton, FL 33487-2742

Printed on acid-free paper

International Standard Book Number-13 978-1-138-63185-4 (Hardback)

Visit the Taylor & Francis Web site at
http://www.taylorandfrancis.com

and the CRC Press Web site at
http://www.crcpress.com

Contents

SECTION II Animaps, 1990s

SECTION III 3D Maps: Georeferencing, Turn of the Millennium

Preface

Spatial Thinking: An Intertwined Pattern of Connected Contributors

There is nothing new about 'spatial thinking'; it has been around as long as people have thought about the Earth. Where do I get food and water for my family and how do I move it from its source to where I want it? Why does the sun move across the sky, and what are the implications for that in terms of raising crops? What is a good way to transport heavy objects across bumpy terrain? How do I get my goods to market? What is an efficient, sensible, and enduring way to communicate across space?

These are all big questions with complicated answers; all those complicated answers possess elements of spatial thinking. They, and millions of others, are enduring questions. Indeed, spatial thinking itself is an enduring process. In this book, we present elements of spatial thinking based on our experiences with the process: both in terms of context and method. As a group, the four principals have over a century of experience in this regard. In some instances, we have been fortunate enough to be forerunners of the tide to come, in others we have served as leaders, and in others we have served as educators and authors. To illustrate our intertwined connection patterns, we offer the following timeline, along with commentary, to illustrate one complicated set of interactions involving elements of spatial thinking. Naturally, others will have different 'twine-lines'; this one is by no means comprehensive or exhaustive. We hope it is suggestive of the importance of understanding the evolution of process. What worked in yesterday's world may no longer be 'trendy'; that does not, however, make it unimportant.

Others, elsewhere, may embrace technique and knowledge that is no longer avant-garde in one locale. Indeed, the past may resurface as one 'rethinks' big issues. Thus, we present a variety of methods from various time periods and these are intertwined, conceptually at least, in various ways. The past offers insight into the present and glimpses of the future: from evolution to revolution?

The following timeline shows the decade leading up to the beginning of the materials in this book followed by the decades of methods used in the materials covered in this book. The decade indicates the time when the method was looked upon as novel and perhaps as a forerunner of things to come. The column adjacent to the decade column lists some of the larger activities of the principals that have helped to shape their approach to the content in this work and elsewhere. The table shows which section of the book is associated with which method/decade and a summary, by title, of the material in that section (note that the environmental material might belong in any decade; it is the method that is captured by decade). We emphasize the importance of a deep understanding of a variety of methods: yesterday's 'clunky' method may transform into tomorrow's 'trendy' method. Today's 'cool' method may become tomorrow's 'old-fashioned' method. It is worthwhile to understand a wide variety of processes and to evaluate their capabilities in a rational and thoughtful manner.

Timeline	**Co-editor roles**
1980s	Forerunners: • Co-creator of TIGER files for US Census (JK) • 1985: Creator and founder, Institute of Mathematical Geography (SA) • Interdisciplinary approaches to teaching about the influence of cultural values on resource use and environmental attitudes (AL) Early adopters: • TeX (SA) • Fractals (SA)
1990s **Section II: Animaps** • Ch. 4: Animaps: Varroa • Ch. 5: Animap Timelines • Ch. 6: Animap Abstraction	Forerunners: • 1990: Creation of very early online, born digital, professional journal, *Solstice: An Electronic Journal of Geography and Mathematics* (SA) • 1999: Community Service Award, awarded by the City of Ann Arbor (SA) • 1990: Created hydrography, elevation, and other digital map data for USGS (JK) • 1995: Researched early adopters of GIS in education (JK) • Awards: Pathfinder Award (1991), DreamKeeper Award (1993), Harold R. Johnson Diversity Service Award (1999) (AL) • LandView Training development for USEPA (MN)
Turn of the millennium **Section III: 3D Maps, Georeferencing** • Ch. 7: 3D Maps: Varroa • Ch. 8: 3D Charts: London • Ch. 9: 3D Charts: Detroit • Ch. 10: 3D Tree Inventory • Ch. 11: 3D Georeferencing: Ann Arbor	• 2002: Co-creation of Wiley's first eBook (SA) • Pirelli INTERNETional Award; finalist, 2002 competition (in top 20, worldwide of over 1200 entries) for *Graph Theory and Geography: an Interactive View* e-book, equal co-authors with William C. Arlinghaus and Frank Harary, John Wiley, New York, April 2002 (SA) • Pirelli INTERNETional Award semifinalist, 2001 competition (in top 80, worldwide, of over 1000 entries) for *Solstice: An Electronic Journal of Geography and Mathematics* (SA) • 2000–2005: Created educational GIS lessons for K-12 and university levels (JK) • Creation of GEOMAT method for organization of maps, archives, and timelines (AL)
2000s **Section IV: GEOMATs** • Ch. 12: Varroa GEOMAT • Ch. 13: Groundwater GEOMAT • Ch. 14: Pacemaker GEOMAT • Ch. 15: Detroit GEOMAT • Ch. 16: GEOMAT Guide	• Pirelli INTERNETional Award semifinalist, 2003 competition (in top 80, worldwide, of over 1000 entries) for a section from a book entitled *Spatial Synthesis*; May 2004, entire e-book published; June 2005, Institute of Mathematical Geography (SA) • National Award: The President's [G. W. Bush] Volunteer Service Award, March 11, 2004 (SA) • 2000–2005: Created educational GIS lessons for K-12 and university levels (JK) • 2008–2011: President and Vice-President National Council for Geographic Education (JK) • Implementation of GEOMAT in the classroom, Ann Larimore with Robert Haug and Sandra Arlinghaus (AL) • Urban Sustainability Directors Network (MN)

(*Continued*)

2010s **Section V: Story Maps** • Ch. 17: Types of Story Maps • Ch. 18: Varroa Story Map • Ch. 19: Tree Canopy Story Map • Ch. 20: Web Maps as Story Telling Ch. 21: In Closing	• 2013: CRC book *Spatial Mathematics: Theory and Practice through Mapping* (SA & JK) • Developed web-based GIS tutorials, 4000 videos, and other supportive materials for the use of GIS in geography, biology, mathematics, business, health, and other fields (JK) • Co-editor of *GIS&T Body of Knowledge* and advisor to GEOTech Center community college model GIS course development (JK) • 2018: TED Talk entitled "The Whys of Where" (JK) • Co-founder Michigan Green Communities network (MN) • Co-founder Great Lakes Climate Adaptation network (MN) • EPA Board of Scientific Counselors (MN) • Presentation of GEOMAT method to various professionals and groups (AL)

Overview of the Book

The content of the book is arranged into sections according to the time periods shown in the table, based on when the particular tool came into widespread use. Thus, in the 1990s, animated maps; at the turn of the millennium, 3D maps; in the 2000s, GEOMATs; and, in the 2010s, story maps. The different chapters exhibit uses of tools based in differing environmental contexts and with differing emphases on maps, archives, and timelines. The reader is presented with ample opportunity, much of it cutting edge at the time, to garner ideas of how one might employ techniques.

To begin, we strongly encourage the reader to read the introductory matter in Section I, preceding the heart of the content beginning in Section II. In it, he/she should gain an appreciation for the broad concepts underlying much of what is to follow. This material introduces the reader, through concept and example, to the underlying philosophy and format for the rest of the book. The intertwined motif suggests the intertwining of chapters that might (superficially) appear to be clearly separated from each other.

To that end, we offer a brief view, tied to the title, of what one might extract, conceptually, from each chapter; we indicate what elements of 'spatial thinking', 'environmental context', and 'maps, archives, and timelines' appear in each chapter. The thoughtful reader will extract more!

Section II focuses on the oldest technique, still evolving and still incorporated in a wide variety of contexts today: the animated map, or 'animap'. In this section, the reader will find spatial thinking concepts associated with diffusion, spatial tracking, space-time patterns, reader/author control of maps, overlays, layering, visualization, georeferencing, and dimensions. He/she will read about case studies in environmental contexts that involve the global honeybee population, the endangerment of a species, disease proliferation, the endangerment of human populations, floodplain issues, and abstract urban construction. The tools involved include elements of maps, archives, and timelines drawn from flat maps and field data.

In Section III, the initial chapter demonstrates the tools of this section (3D maps, which were new to many in the early 2000s) in the same environmental context as in the initial chapter of Section II. The intent is to offer the reader a chance to see the merits and drawbacks of the new tools in a familiar setting. Added elements of spatial thinking in this section center on the rank and size of cities and towns, scale, point distributions, clustering, population–environment dynamics, georeferenced 3D bar charts, interrelations among physical systems, linking 3D mapping with data from the field, geosocial networking, accuracy vs. precision (five arrows clustered inside a bulls-eye of a target are 'accurately' located; five arrows clustered in the outer ring on a target are 'precisely', but not 'accurately', located), and classical concepts from the calculus such as contours as level curves and applications of Archimedes' Principle of Displacement. These concepts are associated with a variety of environmental contexts ranging from urban populations in Europe and North America, air pollution, environmental justice, floodplains, and citizen science. Some of the tools associated with maps, archives, and timelines involve census data (from various locales), Environmental Protection Agency data, site-specific photographic evidence, localized data from neighborhood groups, and elements of virtual reality. In this section, as in Section II, QR codes are employed so that the reader of the printed book has the capability to move directly from a hard-copy document, using a smartphone, to the actual interactive display stored on a remote server. The book comes alive.

Section IV introduces the GEOMAT (Geographic Events Ordering Maps, Archives, and Timelines) method for presenting spatial and temporal materials as a unit. Again, the initial chapter in the section focuses on the same application as the initial chapter in each the two previous sections so that the reader might have the opportunity to compare and contrast differences in method within a familiar context, making the method rather than the context stand out. The reader will find many of the spatial thinking elements of previous sections in this one as well and, in addition, ideas associated with the transfer of technology from developed to developing nations, the use of QR codes to introduce persistence to files archived online, and the enumeration of principles used for GEOMAT formulation. He/she will see them play out in environmental contexts, emphasizing groundwater contamination pollution, environmental justice and public policy, the recycling and reuse of medical devices centered on life-saving applications, and historical urban settings. Maps, archives, and timelines employ field data, 3D maps, QR codes, timeline orientation, and more.

Finally, in Section V, we come down to the present. The first chapter in that section introduces the 'story map'; it does so in advance of presenting the same initial environmental context of the previous sections in case the reader is totally unfamiliar with story maps. Then, the following chapter picks up on the theme of using the same environmental context of the previous three sections so that the reader might once again compare and contrast techniques. Many of the spatial thinking and environmental concepts from previous sections are included here as well. The reader of this section will have ample opportunity to reflect on them and to contemplate how the contemporary tool of the story map might draw from earlier methodologies presented in previous sections, all with an eye to considering where future paths in spatial thinking might lead.

Sandra L. Arlinghaus, Joseph J. Kerski,
Ann Evans Larimore, and Matthew Naud

Acknowledgments

The concepts of 'spatial thinking', 'environmental contexts', 'maps', 'archives', and 'timelines' are not new. Only the first of these is a relatively recent phrase, although certainly the concept of thinking about how space is described, used, and created, and how it meshes with other concepts such as 'time', is not new at all. Indeed, many who have gone before us, from classical antiquity onward, have engaged in 'spatial thinking' in one way or another. They may not, however, have called it that. The word 'spatial' might be thought of as a modifier in various phrases: spatial organization, spatial analysis, spatial statistics, spatial thinking, and so forth. Names such as Abler, Adams, Gould, Upton, Cressie, Griffith, Haining, Goodchild, Janelle, Tobler, and a host of others might spring to mind. The reader interested in delving into the history of all things 'spatial' will find a wealth of materials on the internet. To them all, we owe the great debt that scholars owe to those whose work has come before them.

Many of the materials in this book have their roots in articles that appeared earlier in *Solstice: An Electronic Journal of Geography and Mathematics*. The links in this work have targets, insofar as is currently possible, that are files in persistent archives. To keep sets of figure captions simple, we use acronyms as follows:

Institute of Mathematical Geography, IMaGe. The general site is http://www.imagenet.org from which materials may be retrieved either from a link to a contemporary site or from a persistent archive, Deep Blue, at The University of Michigan (https://deepblue.lib.umich.edu/handle/2027.42/58219). In the caption, only information to retrieve links to the particular image is given; readers wishing to know more should go to the general sites (URLs above). The images on the IMaGe site are original creations first appearing on that site, often using software from Esri, Google, and Adobe software.

Environmental Systems Research Institute, Esri. The general site is http://www.esri.com/

We thank these organizations, and others, for their permissions to include selections here.

In addition, the authors of Chapters 6, 10, and 11 have specific individuals to thank, as their work has a 'citizen science' connection. With citizen participation, be it from students, government officials, or others, it is appropriate to mention them by name to acknowledge their important contributions.

In Chapter 6, the 3D Atlas of Ann Arbor project was a collaborative effort and culminated in three online atlases as well as a variety of other documents (all housed in the persistent Deep Blue). In these atlases, there are tens of thousands of files present, or links to files, to create a variety of images for use in application. New (at that time) files from the city, for extruded structures for the entire city and for contours at a 1 ft. contour interval, opened the door to considering various applications. The primary group of the 3D Atlas team was formed in Engineering 477, Virtual Reality, at the University of Michigan College of Engineering. The course was taught by Professor Klaus-Peter Beier, who invited faculty advisors to work with him and create teams of students to work on projects of specific interest to the advisors. Arlinghaus was one such invitee over a 3-year period, during which she created and led a team to develop the 3D Atlas of Ann Arbor.

3D Atlas Team Composition

- Students:
  - 2003: Taejung Kwon, Adrien Lazzaro, Paul Oppenheim, Aaron Rosenblum
  - 2004: Nikolai Nolan, Rasika Ramesh, Itzhak Shani
  - 2005: Alyssa J. Domzal, Ui Sang Hwang, Kris J. Walters
- Teaching staff, graduate student instructors:
  - 2003: Thana Chirapiwat, Jamie Cope
  - 2004: Thana Chirapiwat, Jamie Cope, Bonnie Bao
  - 2005: Jamie Cope, Brian Walsh
- Other faculty and external advisors:
  - 2003: Matthew Naud, environmental coordinator, City of Ann Arbor; John D. Nystuen, prof. emeritus, University of Michigan
  - 2004: Matthew Naud, environmental coordinator, City of Ann Arbor;
  - 2005: Matthew Naud, environmental coordinator, City of Ann Arbor; Paul Lippens, intern, City of Ann Arbor; Braxton Blake, composer and conductor of music; Edmond Nadler, adjunct prof. of mathematics
- Staff in the 3D Laboratory, Duderstadt Center, University of Michigan: Professor Klaus-Peter Beier, director of the 3D Laboratory; Scott Hamm; Steffen Heise; Brett Lyons; Eric Maslowski; Lars Schumann

Beyond the University of Michigan College of Engineering, there were others with overlapping interests who offered materials, interesting perspectives, and so forth.

- City of Ann Arbor staff: Matthew Naud, Paul Lippens, Wendy Rampson, Chandra Hurd Gochanour, Merle Johnson, Jim Clare
- Members of the Downtown Residential Task Force (DRTF), Downtown Development Authority (DDA), City of Ann Arbor: Susan Pollay, executive director DDA; Fred J. Beal, Jean Carlberg, Robert Gillett, Karen Hart, Douglas Kelbaugh, William D. Kinley, Steve Thorp, Frances Todoro, Wendy Woods; substantial citizen input also from Brian Barrick, Ray Detter, and Peter Pollack
- University of Michigan, Space Information and Planning, Plant Extension: AEC: Donald T. Uchman, coordinator of Space Graphics

Meetings, Early and Mid-2000s

- Ordinance Revisions Committee of City Planning Commission: Preliminary trials, 2000–2003.
- 3D models of various segments of Ann Arbor created for Naud, in response to his specific request, 2003.
- Presentation to Community Systems Foundation involving potential for applications of 3D models in developing nations, October 2003.
- Engineering 477, Virtual Reality, University of Michigan, Professor Klaus-Peter Beier, Fall, 2003: Student team (Kwon, Lazzaro, Oppenheim, and Rosenbaum) developed, with faculty advisors Arlinghaus, Naud, and Nystuen, the first view of a few blocks of downtown with building textures applied. This was an exciting model in that it gave a more realistic view than previous ones. It ran too slowly on common machines to be used by the DDA/DRTF.
- GeoWall display of preceding items, while in the developmental stages, November 11, 2003, at the Duderstadt Center, University of Michigan. City Council and others were invited to offer input at the demonstration.
- Arlinghaus made thousands of files in the 3D Atlas, many of which were offered to those from the DRTF interested in interacting with them. Brief discussions with Pollay about a 3D Ann Arbor (DDA) interactive game.
- Presentation of 3D models (by Arlinghaus), with associated architectural and planning implications commentary by others, using interactive and animated 3D models in Council Chambers at a public hearing hosted by the DDA, April 27, 2004, and written about by Gantert in the *Ann Arbor News*.
- Full set of DRTF models presented by Arlinghaus, in a three-CD set, to the mayor, to each council member, and to the DDA, summer 2004.

- Presentation (SA) to thev Community Systems Foundation involving the potential for applications of 3D models in developing nations, October 2004.
- Engineering 477, as previously, fall 2004. Student team (Nolan, Ramesh, Ishani) developed, with Arlinghaus and Naud, a view of Huron Street with textures. Choice of site to consider came after a suggestion from council member Carlberg and from DRTF member Kelbaugh.
- Presentation, at the invitation of Matt Naud, August 2005, to a group of city staff in the Emergency Operations Center.
- Conversations with Mike Batty, University College London, about the development of Virtual London.
- Participation in teleconference with Calthorpe, September 2005.
- Presentation to Community Systems Foundation involving potential for applications of 3D models in developing nations, October 2005.
- Environmental application of the 3D Atlas, fall 2005, Engineering 477, as previously. Student team (Domzal, Walters, and Hwang) with faculty advisor Arlinghaus with external advisors Naud, Lippens, and Blake. Choice of topic developed in consultation with Naud.
- Ongoing discussions with the 3D Laboratory involving the Virtual Ann Arbor and Google Earth; meeting with Beier and staff (Eric Maslowski, Lars Schumann, Brett Lyons, Brian Walsh, Jamie Cope, Steffen Heise, Scott Hamm, and others), November 2005.
- Discussion with Edmond Nadler, adjunct professor of mathematics, University of Michigan regarding use of mathematical modeling software to create a smooth terrain file that can be extended to the entire city and is also manageable in size.
- Consideration of how to get some of the projects into the immersion CAVE.

In Chapter 10, numerous people were involved directly and indirectly in this project that brought together county and city officials, engineers from a local engineering firm, and members of the public from a variety of neighborhoods including differing residential types and zoning. We thank Harry Sheehan, Greg Marker, Jane Lumm, Roger Rayle, Janice Bobrin, Matthew Naud, William C. Arlinghaus, Edward Goldman, Andrew Nolan, and all the member neighborhoods of the Huron Valley Neighborhood Alliance and the individuals from those neighborhoods who participated in special public and private meetings and in so many other helpful ways.

In Chapter 11 we express appreciation to many who attended various presentations, in the early/mid-2000s, associated with showing files based on the described strategy and for help in creating various forms of them.

- 3D Laboratory, University of Michigan: Klaus-Peter Beier, Lars Schumann, Scott Hamm

- Merle Johnson of the City of Ann Arbor ITS Department and Chandra Hurd (later Gochanour) of the City of Ann Arbor Planning Department
- Donald T. Uchman, Coordinator of Space Graphics, Space Information and Planning, Plant Extension, AEC, University of Michigan
- Current files, and their immediate predecessors, were shown to, or discussed with, various groups to elicit feedback during 2006.
- Community Systems Foundation Annual Conference
- Eric Lipson (vice-chair, City of Ann Arbor Planning Commission) and Vince Caruso (chair, Allen Creek Watershed Group)
- City of Ann Arbor Planning Commission
- Matthew Naud, City of Ann Arbor Environmental Coordinator
- Community Systems Foundation group: John Nystuen, Gwen Nystuen, Fred Goodman, Barton Burkhalter, Ann Larimore
- Board of directors of local League of Women Voters (Shirley Axon, Judith Mich, and others)
- Tracy Davis (*Ann Arbor News*), Vivienne Armentrout (*Ann Arbor Observer*)
- Other groups, including folks from the city and from the university

In terms of direct assistance with this book, the co-editors wish to thank many at CRC Press: our acquisitions editor and team leader, Irma Shagla Britton, senior editor of environmental sciences and engineering, and her staff members; Claudia Kisielewicz, editorial assistant, and, later, Rebecca Pringle, editorial assistant; our project editor, Edward Curtis, and Lara Silva McDonnell and the team at Deanta Global; and Nora Konopka, editorial director of engineering and publisher of environmental science. Their continuing support and wisdom was critical in bringing this project to fruition.

Naturally, our greatest debt of gratitude is to our families, as they watched us spend hundreds of hours working on the development and implementation of this project. Thank you all!

Editors, Co-Editors, and Principal Contributors

Sandra L. Arlinghaus, PhD: Adjunct professor, School for Environment and Sustainability; board member, Project My Heart, Your Heart, Frankel Cardiovascular Center, and Chene Street History Study, Institute for Research on Labor, Employment, and the Economy, both at the University of Michigan; former planning commissioner and environmental commissioner, City of Ann Arbor; former board member, Community Systems Foundation.

Joseph J. Kerski, PhD GISP: Geographer, education manager, Esri, and instructor of GIS, University of Denver and North Park University; former president of the National Council for Geographic Education; former cartographer at the US Geological Survey and geographer at the US Census Bureau.

Ann Evans Larimore, PhD: Emerita professor of geography and women's studies in the Residential College and in the College of Literature, Science, and the Arts, University of Michigan.

Matthew Naud, MS, MPP: EPA Board of Scientific Counselors, vice-chair Sustainable and Healthy Communities Subcommittee; former environmental coordinator and emergency manager, City of Ann Arbor; former member of the Planning Committee, Urban Sustainability Directors Network.

Contributors

Kerry Ard, PhD: At the time of writing of the base document, she was a PhD candidate in the School of Natural Resources and Environment, University of Michigan. Currently assistant professor of environmental sociology, School of Environment and Natural Resources, College of Food, Agricultural, and Environmental Sciences, Ohio State University.

David E. Arlinghaus: At the time of writing of the base document, he was an undergraduate student, Washtenaw Community College, Ypsilanti, Michigan.

Michael Batty, PhD: Bartlett professor of planning and chairman of the Centre for Advanced Spatial Analysis (CASA), University College London.

Thomas C. Crawford, MD: Associate professor, cardiology/electrophysiology, Frankel Cardiovascular Center; director, Project My Heart, Your Heart, Frankel Cardiovascular Center; both at the University of Michigan.

Kim A. Eagle, MD: Albion Walter Hewlett professor of internal medicine, professor of health management and policy at the School of Public Health, and director of the Frankel Cardiovascular Center; co-founder, Project My Heart, Your Heart, Frankel Cardiovascular Center; all at the University of Michigan.

Salma Haidar, MPH, PhD: At the time of writing of the base document, she was working on her PhD dissertation under the direction of Mark Wilson, University of Michigan School of Public Health. Currently she is associate professor, Division of Public Health, School of Health Sciences, Central Michigan University.

Robert J. Haug, PhD: At the time of writing of the base document, he was a PhD candidate in the Department of Near Eastern Studies, University of Michigan. Currently he is associate professor in the

Department of History and director of the program in Middle Eastern Studies at the University of Cincinnati.

Marian Krzyzowski: Director, Chene Street History Study, Institute for Research on Labor, Employment, and the Economy, University of Michigan; lecturer, Frankel Center for Judaic Studies, University of Michigan.

Karen Majewska, PhD: Chene Street History Study, Institute for Labor, Employment, and the Economy, University of Michigan; mayor of Hamtramck, Michigan.

John D. Nystuen, PhD: Professor emeritus, Taubman College of Architecture and Urban Planning, University of Michigan; retired chief executive officer, Community Systems Foundation.

Roger Rayle, MSE: Leader, Scio Residents for Safe Water; venture catalyst, Roger Rayle Virtual Ventures LLC (R2VIVE).

Diana Sammataro, PhD: Retired research entomologist, USDA-ARS Carl Hayden Honey Bee Research Center, Tucson, Arizona; currently DianaBrand Honey Bee Research, LLC.

Mark L. Wilson, ScD: Professor Emeritus, Department of Epidemiology (School of Public Health), University of Michigan; Department of Ecology and Evolutionary Biology (College of Literature, Science and the Arts), University of Michigan.

I

Introductory Matter

The section introduces the reader, through concept and example, to the underlying philosophy and format for the rest of the book. The intertwined vine motif suggests the intertwining of chapters that might (superficially) appear to be clearly separated from each other.

1

Spatial Thinking: Maps, the New Paradigm

Joseph J. Kerski

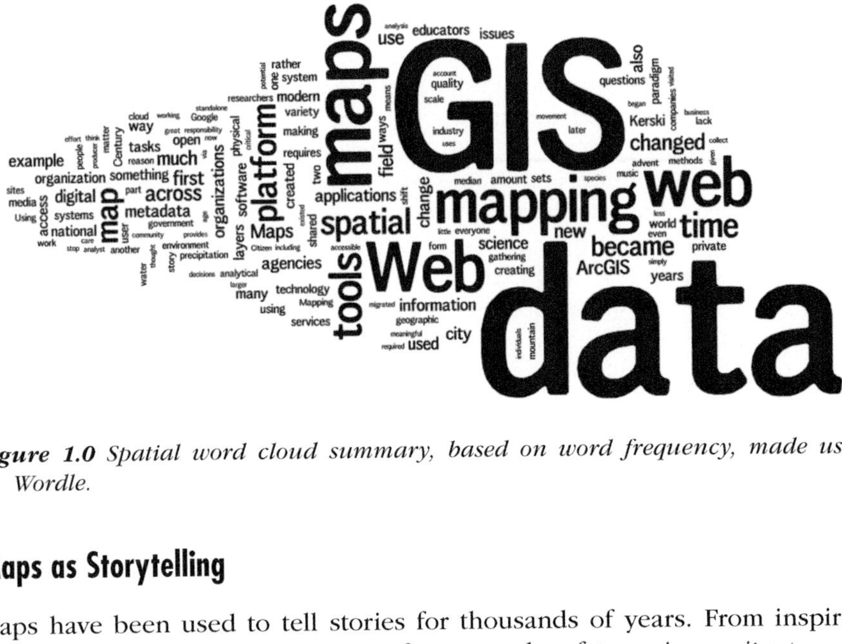

Figure 1.0 *Spatial word cloud summary, based on word frequency, made using Wordle.*

Maps as Storytelling

Maps have been used to tell stories for thousands of years. From inspiring exploration through the depiction of vast swaths of *terra incognita*, to cultivating a sense of unity for the new United States in the late 1700s, to maps of the solar system during the initial Space Age of the 1960s, maps have been effective storytelling tools for several reasons. One key reason is because maps have proven to be an efficient means of conveying a large amount of information in a small amount of space. Another reason is the sense of awe they have inspired of showing real places, with the map reader seemingly "above it all" with a birds-eye view.

As the centuries passed, as rich as those maps that were scrawled in the dirt, etched on stone, carved in wood, or drawn on paper, linen, vellum, and film have been, modern digital maps are already showing themselves to be even more effective storytelling tools. These maps are created with geographic information systems (GIS). The analytical tools, cartographic elements, spatial data, and multimedia in a GIS allow a greater variety and amount of information to be conveyed in map form than analog methods ever could. GIS allows the map reader and spatial analyst to derive information about the place, region, issues, and phenomena under study. It also allows the publishing of mapped information into an ever-expanding variety of formats, reaching a much greater audience than physical maps ever could. Furthermore, maps are no longer only published by national mapping agencies, statistical agencies, and private mapping companies. With the advent of the citizen scientist movement and crowdsourcing mapping tools, anyone is now a potential map publisher.

Modern GIS

As the first decade of the 21st century came to a close, it became obvious that GIS had changed in fundamental ways during what can now be considered to be its first phase, from 1965 to 2010. Over those first 45 years, GIS tools evolved from working on mainframe computers to minicomputers, and then to PCs, laptops, tablets, and smartphones. Over that same time period, spatial data were housed on physical media. While that media changed from magnetic tape to floppy disk and later to CDs and DVDs, the physical media was at first a godsend to spatial analysts, who were able to send data among systems, organizations, and data users. As time passed, and more users needed access to data, and as the data sets themselves became larger in size and finer in resolution, the constraints of housing data on physical media became apparent. Also, during its first years, GIS was in large part focused on creating mapping products: Instead of using manual methods, GIS enabled maps to be created using digital methods. The maps could then be updated more frequently and in new ways, but GIS in part replicated what had been done before—making printed maps. This was an understandable role for GIS during a time when very little digital spatial data existed. But as time went on, GIS became a 'transformational' technology—going beyond the making of maps to a visualization and decision-making tool. A similar thing occurred with the advent of motion pictures. When motion pictures first began, they were primarily used for filming stage plays. They were used as a new technology to enhance an existing medium. But when the movie cameras were taken outside, motion pictures became their own industry, their own art form, their own social movement, separated from stage plays. GIS is now its own industry, science (geographic information science), and a societal force unto its own; it has become transformational.

During that 45-year time period, the workflows involved in most GIS operations followed the same formula: (1) ask a geographic question, (2) gather the data, (3) analyze the data, (4) make decisions based on the analysis.

For decades, step (2), the gathering of the data, required a great deal of the analyst's time, because the "gathering" included searching for data, which itself was an arduous task given the lack of open data, the lack of metadata standards, and the lack of easy-to-use portals. The gathering also included obtaining the data, which had to be done by obtaining physical media, and later by seeking permission for access, and then downloading. Whichever way the data was gathered, once obtained, the data had to be processed in order for it to be used in a GIS environment. This process included reading and/or converting the data into a geodatabase of some sort (from coverages and grids in the early days and later to shapefiles to geodatabases and other formats), casting the data onto a suitable and appropriate map projection, renaming the field names into something understandable and workable, changing the classification method and symbology, populating the metadata, joining the tabular data with additional layers, and other associated tasks. These tasks were typically very time-consuming, and given the looming deadlines present in any project, they resulted in more time spent on step (2) and consequently less time spent on the analysis step (3).

All of this began to change in 2010. Data became increasingly accessible as stored files—first on FTP sites, later on standard web sites, and then assembled into data portals, and more recently, increasingly as web services. GIS software and tools followed suit—they themselves became available in a software-as-a-service (SaaS) model, accessible via a web browser, anywhere, anytime, and on any device. Even 'desktop' software such as ArcGIS from Esri was no longer truly a standalone entity but became connected to the web. Many functions, such as geocoding, were processed online on the server rather than on the client's own computer. This allowed for faster processing time and the ability to work with larger data sets. Basemaps and thematic maps became available as online layers, rather than as something that one had to download, format, and use on a local device. Workflows and models also became something that the analyst could save and share with another user, so they did not have to recreate the procedures, but could use an existing model, substituting their own data or adjusting some of the model's parameters. These developments led to what is known as *Web GIS*.

Web GIS

Modern GIS is a part of a much larger change in computing—the shift from desktop-only tools to Web GIS. Within GIS, this shift involves more than the technology or tools, but, in addition, includes

- Software products platforms and APIs
- Client/server web services and apps
- Standalone desktop connected devices
- Print maps to maps and data visualizations

- Static data data services, streams, and big data
- Custom applications interoperable packages and libraries
- Single all-purpose applications many pathways and focused apps
- Proprietary data open data and shared services

If all this seems like mere semantics, Web GIS should matter to researchers and educators alike. The reason it should matter to researchers is because it enables researchers in almost any field to ask the 'whys of where' questions using spatial data and GIS without having to set their core field(s) aside and spend a number of years becoming a GIS analyst. The learning curve to access these tools is much reduced. For example, a researcher can access a Web GIS such as Community Analyst or Insights and be up and running quickly, exploring their own data. They can take the results to Tableau, R, Stata, SAS, or another package of their choice. The reason Web GIS should matter to educators is because educators care deeply about meaningful student learning with geotechnologies. To foster spatial and critical thinking with geotechnologies requires more than looking up place names on a map, or routes from a certain point to another point. It requires educators to be purposeful about using maps as the analytical, exploratory tools that they are. It requires that educators cultivate geographic inquiry, critical thinking, and meaningful use of technology.

Maps on the Web versus Web Mapping

Beginning with MapQuest in the 1990s, Yahoo, Google Maps and Google Earth in the 2000s, and the entrance of maps via mobile devices into everyday experience in the 2010s, interacting with maps in digital form rather than paper has become more common. However, 'web mapping' is not the same as simply using maps on the web, no matter what the discipline: health, energy, city planning, or, as is the focus in this book, research and education.

Maps on the Web

Using maps on the web includes looking up a place name, examining thematic maps such as ocean currents, world biomes, or demographic characteristics by neighborhood across a city, finding the distance between two points on a map, finding the route between two points, mapping locations that you have visited in the field, and so on. Certainly nothing is wrong or inadequate with any of those tasks. Using maps on the web focuses on the "What's where?" question.

Web Mapping

By contrast, the focus of web mapping is on examining patterns, relationships, and trends. Web mapping examines changes over space and time at a variety of scales and across themes. For example, what is the relationship between the location of mines and water quality across a mountain watershed, or between median age and median income across a city? How does the land use change across a region over time, or the precipitation across a mountain range? All of these questions end with "And why?" Gritzner (2002) describes geography as being about "What is where, why there, and why care?" Reflecting this, web mapping focuses on the "Why there?" and "Why care?" part of his framework (Table 1.1).

Using maps on the web is a stepping stone to web mapping. While there is an overlap between them, these two processes are not synonymous. Web mapping uses the concept of GIS as a platform, including web, mobile, and desktop, with its analytical, multimedia, and application abilities, to its full potential.

The educational implications of this are many. How do educators teach in this new paradigm of web mapping, and how *should* they teach? What concepts should they teach, and what skills should they seek to foster? What tools and data sets should they use? How should they incorporate new field techniques and applications? How should they assess student work given the ease of creating web mapping applications such as story maps (Kerski, 2018)? How should primary, secondary, community college, and university courses and programs change to encompass this new Web GIS world and provide meaningful skills and visionary perspectives for their students?

In addition, Web GIS is not just "more and better" GIS, it also requires new ways of managing GIS. Like other software that has migrated to the cloud, such as Google Docs, Salesforce, or Dropbox, GIS is accessed via a named user. A named user is logged in to a mapping platform such as ArcGIS Online,

Table 1.1 Using Maps on the Web vs. Web Mapping

Maps on the Web	Web Mapping
Tasks: Navigation Visualization	**Tasks:** Navigation Visualization Analysis Creating web mapping applications Collecting and exploring field-collected data
Sample questions:	**Sample questions:**
Where are the field sites I visited?	Why does the water quality vary across the field sites I visited?
Where are the younger and less affluent neighborhoods in this city?	Is there a spatial and attribute relationship between median age and median income in this city, and if so, what is the relationship, why does it exist, and does it change over time?
Where is the mountain range in a region and what is the precipitation regime across it?	How and why does the precipitation regime change across the mountain range?

and accesses an account on that platform. He or she has roles and access to tools as assigned by the administrator of that online account. He or she uses that account to store data layers, images, tables, maps, and applications, such as story maps. A modern GIS data user might be working with dozens or even hundreds of map layers simultaneously. Data management has always been key to the successful use of GIS, because GIS has always, by definition, been a system with many "moving parts."

Web GIS is not simply "doing GIS tasks on the web" that formerly were performed on standalone workstations, but rather, has altered the way organizations work internally and with other organizations. Desktop-only GIS could be thought of as a system of 'records', with each organization maintaining their own data. For example, consider a typical city's separate GIS and non-spatial databases for water, sewage, electrical, gas, zoning, parks and recreation, parcels, streets, and so on. Web GIS can aptly be thought of as a system of 'systems'. It can also be thought of as a system of 'engagement', because it has the potential—indeed, ample evidence already exists (Ganapati, 2010)—to engage citizens and stakeholders in their own organization in a participatory manner that can improve their communities and the lives of people who live in them.

The migration of software and data to the web has changed much more than how and why GIS is used. It has changed the face of just about every industry and the daily work tasks of just about everyone in modern society (Kerski, 2016). As music has migrated to the cloud, it has altered how consumers think about how to store and access music, how artists think about disseminating music, and how publishers think about copyrighting and marketing music. As word processed documents, spreadsheets, and photographs have migrated to the cloud, it not only allows people to have more storage capacity, but much more importantly, it has opened up the means by which researchers can collaborate on projects. Similarly, the migration of Salesforce and other business systems has changed the way organizations do business.

Why Web GIS Is Revolutionary

The advent of GIS to the cloud may have occurred more quietly than these other systems, but it is even more revolutionary for several reasons. First, the issues and problems that the GIS community seeks to tackle do not stop at political boundaries: Water quality and quantity, food security, human health, soil erosion, energy, urbanization, natural resource conservation, sustainable agriculture, sustainable tourism, economic inequality, biodiversity loss, natural hazards, political instability, crime, and other concerns of the modern world do not stop at city, state, national, precinct, or any other boundary.

Second, neither do these issues stop at disciplinary boundaries; these issues, and all the other major issues of our 21st century world, transcend such disciplines as geography, engineering, design, sociology, biology, hydrology, geology, planning, mathematics, history, language arts, anthropology, business, economics, and many more.

Third, the complexity and severity of these issues—many of them life-threatening—*requires* intergovernmental and interdisciplinary cooperation and collaboration. Web GIS fosters cooperation and allows collaboration in a way that the former GIS paradigm could not. The GIS paradigm of the first 45 years was not only arduous, but it was by its nature an individual effort (or at best that of a small team). Each person worked on his or her own workstation on his or her own project; the size of the data sets made them difficult to share, and few collaborative tools existed for even analysts in a city government, for example, to simultaneously collect, edit, and analyze data in their own departments, much less collaborate with other departments or non-government organizations, non-profits, academia, and private industry. Web GIS changes all of that because not only can data be shared, but models, tasks, tools, and workflows can be shared. Web GIS is therefore not simply "doing GIS on the web" but represents a paradigm shift for GIS.

Fourth, web GIS has changed every data user to a potential data 'producer'. For centuries, individual cartographers such as Mercator, Varenius, Al Idrisi, and others, labored for years, sometimes decades, making a single edition of a map. With the arrival of the printing press, private companies could print multiple copies of maps, and these documents slowly made their way into the hands of the public. During the 19th century, the rise of national mapping agencies such as the Ordnance Survey and the US Geological Survey were tasked with mapping their entire countries at large scale, resulting in tens of thousands of map titles. The rise of private mapping and satellite image companies such as Klett-Perthes, Rand McNally, Nystrom, Esri, DigitalGlobe, and others expanded the types and scales of maps, and, along with technical changes in the national mapping agencies, ushered in the age of digital map production. Yet for all of these changes, the art and science of creating maps and digital data rested in the hands of relatively few people. Two things changed this paradigm.

Implications of Web GIS

Web GIS was the first development that changed the face of traditional GIS. The advent of data and maps as services meant that GIS became a platform that people could build on. With Web GIS, anyone can publish data onto a mapping platform, such as ArcGIS Online from Esri. The platform is completely open to government agencies, private companies, non-profit organizations, educational institutions, and individuals, who may consume data *from* it but also produce data *for* it. This means that the same platform contains data from the United Nations Environment Program to a middle school student working on an assignment, and everything in between. A minimal amount of metadata is required for each map published, as well as a log-in, but other than those two requirements, the platform is completely open.

Citizen science was the second development that changed mapping. Citizen science, the act of creating scientific data from non-scientist citizens, is nothing new; indeed, the birding community has been collecting detailed data about species, sounds, feather color, tree species, time of day, date, season, behavior, and other information about birds since the late 19th century, enabling the creation of such organizations as the Audubon Society. During the 20th century, other citizen groups began to collect data on weather, phenology, animals, water quality, and other information. One of the earliest widespread examples was the GLOBE program (Global Learning and Observations to Benefit the Environment; www.globe.gov), an international science and education program that provides students and the public around the world with the opportunity to participate in data collection and the scientific process. GLOBE data collection has focused on soil conditions, temperature, and precipitation. Another example is OpenStreetMap, a global effort by individuals to create a digital street atlas, particularly in areas where no national mapping agency existed, for use by government agencies and also by individuals. Still another example is Mapillary, an effort to generate Google Streetview-like imagery for mapping and inventorying light poles, streetscapes, trees, and other natural and human-built objects, in areas not covered by Google's vehicles and cameras.

Web GIS tools have enabled this 'crowdsourced' data to be easily mapped and used in spatial analysis. Through tools such as iNaturalist, Collector for ArcGIS, and Survey123, field-collected data can be entered into a form or placed on a map, and the results run through spatial statistics and visualized on 2D or 3D maps, including story maps (Kerski, 2017). Collected data ranges from the conditions of sidewalks, invasive species, pedestrian counts, shaking from earthquakes, litter and graffiti, and much more (Kerski, 2015). Some data is actively sought by local and national government agencies that lack field staff to collect the data themselves. Some of these tools, such as Collector for ArcGIS and Survey123, create editable web mapping layers inside the ArcGIS Online platform. These layers are combined in many ways to make everyday decisions from the local to global scale.

Citizen science and Web GIS as a platform has numerous societal implications. One is data quality. Because the platform is open to any publisher, the data consumer needs to cultivate the habit of being critical of the data, asking such questions as "Who created this data?"; "Can I trust it for my purposes?"; "How often is the data curated or updated?"; "What methods were used to create the data?"; "Why was the data created?"; and "At what scale was the data created?" A minimal amount of metadata is required by most platforms for data to be published, but much data is published with very little metadata. How can one make a wise decision when one encounters a data set containing so little metadata? Thanks to citizen science, Web GIS, the open data movement, and agreed metadata standards, spatial data has never been as rich in scale, variety, and quantity as it is today. But with great opportunity comes great responsibility: It is the responsibility of the data producer to provide 'truth in labeling' (metadata), and it is the responsibility of the data consumer to determine 'fitness for use' (Kerski and Clark, 2012). Increasingly, almost everyone involved with GIS is both a data producer and a data consumer.

The GIS Platform

The modern GIS environment is a platform. By definition, a platform is something that can be built on, with specific applications for specific needs. This too represents a shift for GIS because, formerly, geospatial technology was so specialized and 'niche' that in most workplaces, GIS analysts were housed in a special area, and if someone needed a map or spatial data, they would go talk with the GIS staff. Nowadays, GIS is increasingly seen as an enterprise asset, something that is valued, funded, and supported throughout the organization, and something that every employee in the organization should have access to and at least be familiar with. Story maps are easily accessible to everyone in an organization and thus are becoming one means by which everyone in that organization works with spatial data.

ArcGIS is an example of such a modern GIS environment. ArcGIS provides a platform for spatial data, maps, and tools to be created, modified, analyzed, embedded, searched, and shared. A GIS allows for an entire geographic workflow to be accomplished—from gathering the geospatial data, to performing analytical techniques on the data, making decisions and evaluations based on the data and techniques, and finally creating output from that data.

GIS is one of the few sets of computing software that allows a problem to be taken from its conceptualization all the way through to its communication. A problem can be examined, represented, analyzed, modeled, and communicated within a GIS environment, making it suitable and valuable for problem-solving and decision-making. GIS output includes an expanding variety of multimedia that can be shared in a number of ways—including story maps.

References

Ganapati, S. 2010. Using geographic information systems to increase citizen engagement. *The Business of Government*: Spring.

Gritzner, C. F. 2002. What is where, why there, and why care? *Journal of Geography* 101(1): 38–40.

Kerski, J. J. 2015. Geo-awareness, geo-enablement, geotechnologies, citizen science, and storytelling: Geography on the world stage. *Geography Compass* 9(1): 14–26.

Kerski, J. J. 2016. *Interpreting Our World: 100 Discoveries that Revolutionized Geography*. Santa Barbara, CA: ABC Clio, 386 p.

Kerski, J. J. 2017. Connecting citizen science, GIS, community partnerships, and education. Citizen Science GIS Blog, University of Central Florida. http://www.citizensciencegis.org/connecting-citizen-science-gis-community-partnerships-and-education-a-guest-blog-by-dr-joseph-kerski-of-esri/.

Kerski, J. J. 2018. Why GIS in education matters. Geospatial World, blog, August 6. https://www.geospatialworld.net/blogs/why-gis-in-education-matters/.

Kerski, J. J., and J. Clark. 2012. *The GIS Guide to Public Domain Data*. Redlands, CA: Esri Press.

2

Spatial Thinking: Archives and Timelines

Sandra L. Arlinghaus

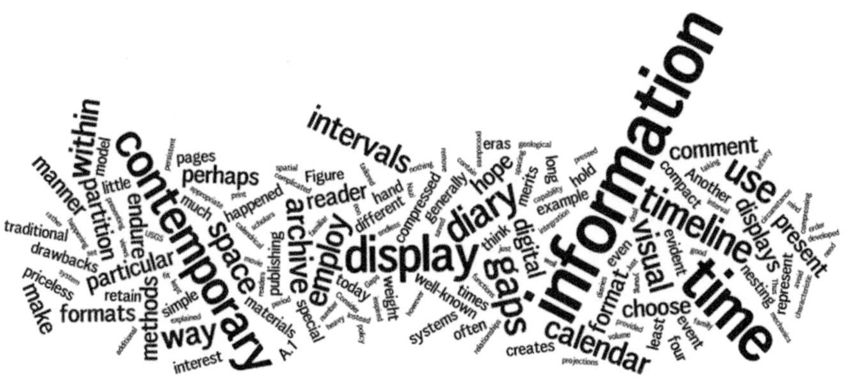

Figure 2.0 Spatial word cloud summary, based on word frequency, made using Wordle.

For centuries, scholars and people from all different backgrounds have developed methods and tools for organizing events in space and time and saved them in contemporary formats that they hope will endure as an archive of materials for future generations to draw from. Some are well known to us today; others may have been well known in earlier times, only to have vanished from view today. One way to avoid the vanishing information phenomenon might be to preserve documents in a variety of contemporary formats in the hope that at least one of them will endure and perhaps prevail.

In the previous chapter, Kerski considered the evolution of mapping and geographic information systems (GIS). Here, we look at a somewhat parallel view of timelines with an eye to considering how they might evolve along with maps. A simple way to organize information is to use a calendar. Some of us use a wall calendar; others might use a desk calendar. Yet others use apps on smartphones. Independent of the mechanics, they all represent the same idea: an even partition of time to which information can be attached. One variant of the simple calendar is the diary. All of these employ a 'calendrical' timeline: a timeline with no collapsed intervals. Gaps of inactivity are as evident as busy times. Simple calendars appear to hold no particular fascination; diaries, on the other hand, might hold a great deal. Consider the diary kept by a young girl. One diary might contain the day-by-day experiences of a girl in an American high school, with particular highlights associated with pages devoted to a special dance event, perhaps a corsage pressed between the pages and comments about her date for the event. To her the diary is very special, indeed priceless, but it might be of little interest to others because it reads so much like so many others. Another girl might have written about her plight in escaping Nazi Germany; her diary not only became priceless to her family but also to much of the world (Frank, 1947). Indeed, the original Dutch document was translated into English (1952), cast as a play (1955), and inspired a movie (1959). The integration of spatial and temporal information compressed into an archive (paper, video, or otherwise) can be powerful.

All of these represent an integration of space and time and do so in a manner that creates, and perhaps draws from, an archive of maps and timelines. The time partition employed is linear in nature; it has even spacing of intervals with no weight given by the partition itself to one interval or another.

One can, of course, choose to weight a timeline. A timeline with varying lengths of intervals, often with compressed intervals where there is little information, is one way to make a compact display. Another is to retain the gaps and, instead, nest information. Geological timelines often employ this latter strategy. Figure 2.1 shows adjacent, vertical timelines: eon, era, system, and epoch. There is nesting of information about eras within eons, of systems within eras, and of epochs within systems. It is an efficient visual display of complicated relationships taking place over a long period of time.

With timelines, it is generally desirable to retain the gaps and not remove them; surely something is happening in the gaps. Often, however, in the interest of conserving space or compressing information for visual appeal or capture, the gaps are removed. Just because the creator of the timeline doesn't know what happened in a particular gap does not mean that nothing happened. Instead, evident gaps in timelines suggest opportunities for additional research!

In this work, we employ four different contemporary techniques for integrating space and time in a manner that creates a visual archive. Each has merits and drawbacks. There are, no doubt, an endless number of ways to create such displays; that is a 'good' problem (just as it is with the infinity of map projections from which to choose). No one method or model is 'perfect' or 'precise'. We choose those methods with which we are familiar

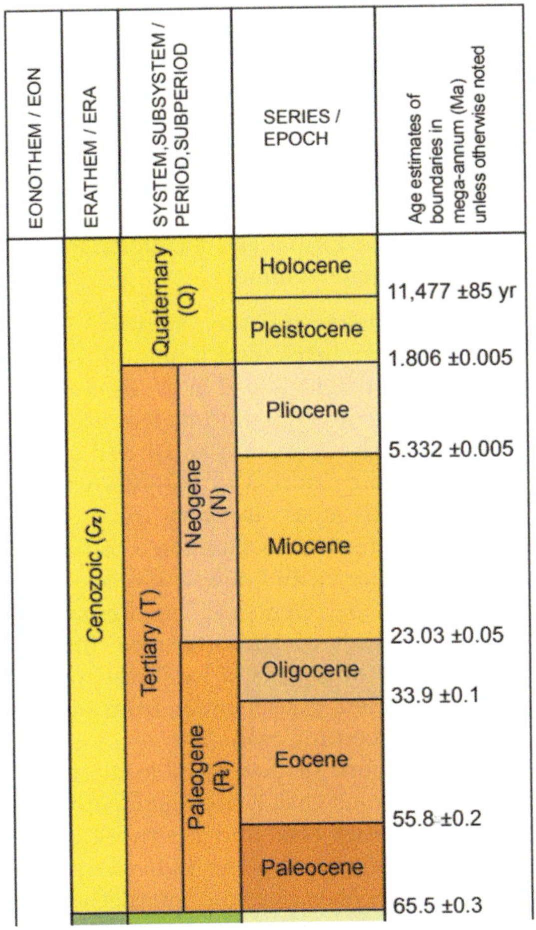

Figure 2.1 *A section of a geological time scale employing the nesting of timelines in order to present information over a long time span in a compact visual manner. (Source: USGS (2006), Divisions of geologic time: Major chronostratigraphic and geochronologic units.)*

and comment for the reader on how the display functions as well as on its merits and drawbacks. For example, one important feature of these displays might be to inform public process or policy. Another might be associated with citizen science. When those characteristics are present, we comment on them.

We begin by presenting, in the next chapter, an introduction in which the four methods of blending space and time, through maps, archives, and timelines, are explained generally with an accompanying example or two. In the core of the book, we display more extensive examples derived from appropriately selected portions of our own works and those of others, all set within various environmental contexts. We comment on the differences and similarities from display to display. To aid the reader in implementation, we present general guidance and specific instruction (as we think appropriate). The contemporary procedures make heavy use of the internet and of the capability of computer technology. Thus, readers are presented with 'snapshot views' of digital materials and are provided with QR codes and links to servers (some persistent, some not) where full digital files are stored. We employ the print format of traditional publishing as one format and the digital format of contemporary publishing, integrated in a single volume.

The two overview chapters, taken together, suggest the persistence of studies that involve all of maps, archives, and timelines as one way to integrate space and time. Where might this idea lead in the future? Of course, one never knows, but perhaps there are directional hints that are available in the study of the past and present. Today, even in a small city embedded in the rural Deep South of the United States, access to the internet is as straightforward as it is in larger metropolitan areas. Delivery trucks bring goods from all over the world to homes there, with a guarantee of two-day delivery. Items previously only available in stories and newscasts have become a reality. The need to go, in person, to department and big-box stores has diminished. Smartphones enabled with apps of all sorts make life easier: load an app and open the menu from any of dozens of local restaurants and order a dish from one restaurant while your guests order from other local restaurants. Pay with a credit card. No need for cash. Throw an elaborate in-home dinner party, using the 'food court' of dozens of local eateries, with no effort from the host. Travel has become easier, both nationally and locally (with the advent of smartphone technology linked to a network of local drivers providing on-demand services). The need for an individual car/vehicle is diminished; ordering goods and services over the internet with a smartphone has become quick and efficient. What might be the short-run implications? Will delivery people's jobs remain stable? In the immediate future, one might see a need for an increased supply. In a more distant time, however, as drone or related technology, such as autonomous vehicles, becomes viable for deliveries of all sorts, delivery jobs might diminish in number. How will the nature of municipal planning change? Road networks might shift to accommodate a vehicular network centered on delivery vehicles. Then again, road networks might eventually become a thing of the past as drone delivery takes over. As big-box commercial establishments continue to be phased out, large parcels of land might become available for extra development of other sorts, including (in addition to uses such as housing and commercial) areas for parks and surface and ground water management or for farming local produce that could be ordered direct from the farm and delivered by drones to individual homes, on demand. Mapping of municipal networks is changing dramatically as web mapping and cloud computing become dominant. Drones are also used as data gatherers. For example, Naud notes that software called 'Drone Deploy' can be used to inspect landfills efficiently and in a small amount of time. Aboveground analysis is easily visible; belowground is also important. Cities might need to learn to create air traffic control maps of 3D space in support of an expanding fleet of drones, as physical 'cloud-type' features, with various categories of drone zones. Zoning and planning commissions might find a whole new set of ways of thinking about life on our planet; the hints available from the past and the present suggest that we are in a transitional stage that is already leading to a bold and exciting new future, with a leading edge visible in the web mapping community.

With this, or another, sort of big-picture context in mind, the reader will hopefully be stimulated to appreciate the importance of spatial thinking in past and present environmental contexts. From that stable and grounded platform, we encourage him/her to leap into the present/future interface to create exciting new applications of spatial thinking, bearing in mind that apps need to be tailored to the situation at hand; make the model fit the circumstance, not the other way around!

References

Frank, A. 1947. *Het Achterhuis: Dagboekbrieven 14 Juni 1942–1 Augustus 1944 (The Annex: Diary Notes 14 June 1942–1 August 1944).* Amsterdam, the Netherlands: Contact.

Frank, A. 1952. *Anne Frank: The Diary of a Young Girl.* New York: Doubleday.

Goodrich, F., and A. Hackett. 1955. *The Diary of Anne Frank* (play).

Goodrich, F. and A. Hackett. 1959. *The Diary of Anne Frank* (screen adaptation).

United States Geological Survey (USGS). 2006. Divisions of geologic time: Major chronostratigraphic and geochronologic units. Retrieved from https://pubs.usgs.gov/fs/2007/3015/fs2007-3015.pdf.

3

Spatial Thinking: Book Structure

Sandra L. Arlinghaus

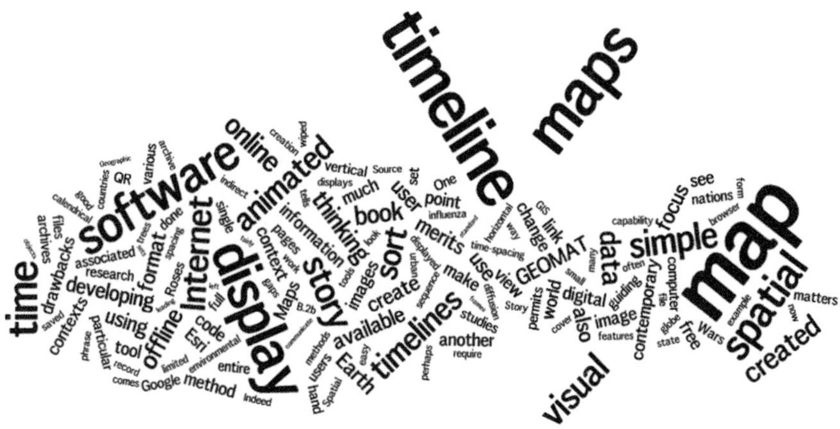

Figure 3.0 Spatial word cloud summary, based on word frequency, made using Wordle.

The Importance of Spatial Thinking

The phrase 'spatial thinking' refers simply to the idea that matters concerning space, the patterns of places, objects, or people on the surface of the Earth, are not only interesting but also critical to guiding a variety of avenues of research and daily life. Thinking about such patterns, at the outset, can guide results. Far too often, maps and other images are seen simply as a way to make a manuscript look nice or perhaps to add a bit of clarity to a table of data. The 'spatial' part is an afterthought. We insist, as do many others in the contemporary world of spatial science, that maps and images of various sorts have a primary role in guiding and leading, not merely in following, research. This work is devoted to providing examples of such, particularly in environmental contexts.

Rationale for Hard Copy

The subtitle of this work, "Maps, Archives, and Timelines," enumerates particular tools that cover different dimensions of scholarly activity. Indeed, one might wonder why a set of co-authors well grounded in geography would even consider creating a book on 'spatial thinking' in conventional print format. The answer is simple: Recall the phrase 'dark ages'. It is a phrase that is out of vogue in contemporary academic circles, but it is one that packs a powerful punch: loosely speaking, it refers to a period in which there is a paucity of recorded information. The key here is 'recorded'. It is very tempting today to think that the wonderful digital files we have, backed up in various formats and in various locales, will be perpetuated into the distant future. But is that the case? No one can be sure.

Look to the environment as an example. Recall beautiful tree-lined streets with elm trees gracefully shading the sun from the concrete as they cooled the yards of adjacent houses. Where are those trees now? Too much reliance on a single species obliterated the record; Dutch elm disease wiped the slate clean. Now, hopefully, we see mixed-species plantings of urban street trees. If one fails, the urban tree record is not wiped out. So too with documents. Total digitalization is nice, but what if that slate gets wiped clean? Do we fall into a 'dark age' with little record of what happened? Thus, this book, as a printed document, is itself an argument for capturing digital files in a form where preservation over time is better known, such as hard copy. And in making the transformation from digital to paper and back again, it is helpful to have hooks to hang onto; hence, we adopt the three axes of maps, archives, and timelines to capture that transformation to aid in the projection from one format to another.

Organization of the Work

The set of sample studies that follow all employ some digital tool that permits the simultaneous capture of maps, archives, and timelines. The four tools, or perhaps more appropriately 'methods', are animated maps, online 3D maps (such as Google Earth), GEOMATs (Geographic Events Ordering Maps, Archives, and Timelines), and story maps (primarily Esri). They are arranged into four sections, one for each method. Each section contains one environmental context that is constant across sections (the honeybee mite example) for the purposes of systematically comparing the methods. Each also contains variety in environmental application to motivate the reader. Finally, where appropriate, sections contain conceptual material about the method. For example, the section on GEOMATs contains a guide at the end of the section to give the reader an overview of the deeper issues associated with this method once he or she has had the opportunity to learn, by example, about GEOMATs. The section on story maps begins with a general description of story map types, then exhibits a couple of studies based on environmental data in subsequent chapters that refer back to the initial one in that section. The section concludes with 'hands-on' materials with which the reader can put into practice what he or she has been reading. Thus, the thoughtful reader should look for connections within sections and between sections; the content is complex and so is the set of connecting ideas that can emerge.

Four Methods

We first give a brief discussion of each method, then follow that with a set of studies in which the reader learns more, through example, about the utility of these methods in leading research questions and guiding the research direction. Some of the examples also offer a vantage point into seeing how the methods have had, or might have, more pragmatic applications in guiding or suggesting directions to municipal authorities in the creation or implementation of public policy. No one strategy is 'best': each has merits and drawbacks. Let the scholar/researcher choose a method that is a good fit to the project at hand.

Animated Maps

A simple animated map shows a succession of images that come into view one after another. One might think of flipping through the pages of an atlas as an animated map, although generally one thinks of them as digital files. These displays are most effective when the map boundaries are the same from image to image with only small changes from one page, or frame, to the next so that the viewer can grasp each change before the next image comes into view. When the sheaf of pages or frames is arranged to cover an expanded sequence of time, an indirect archive is created. Numerous examples of such maps are available on the internet; one that readers might enjoy, if they have not already seen it, is a NASA (2013) video on global temperature variation that shows 'heat' maps of changing global temperatures over a long period of time.

When the time-spacing between successive pages or frames is set at creator-specified intervals, an indirect timeline is inserted. In this method, the maps are clearly visible; archive and timeline are inferred by the time span covered by the entire range of maps and the time-spacing between successive maps. This sort of display can be created online or offline, using free or commonly used software.

When the animation is saved in 'movie' format, the reader can control the time-spacing. The creator will need to decide whether such user control is desirable. In some instances, such as for offering more time to study a particular map or to choose to jump ahead in the sequence, user control may be desired.

On the other hand, if a particular point is being made with the spacing of the timing, then such user interference may destroy that point, and it may be better to present the animation in a fixed format, such as an animated .gif. This method might be viewed as relatively low-tech; it is, however, a method readily available to all levels of users and to users for whom their connect-time with the internet is precious.

The merits to this sort of visual display are as follows:

- The map may show all countries on the Earth in a single frame.
- The map visually emphasizes differences, as these pop out as one moves from frame to frame.
- When needed, the map may be supplemented with an inset map to show small countries that become obscured on a global map.
- The file may be created totally offline and uploaded, in a short amount of time, to a server that distributes files across the internet.
- The maps may be created in GIS software or in simple base maps, available for free. The countries may be colored using standard painting software on the computer.
- Because only simple software is used, it is generally straightforward to update maps as new information comes in.
- Individuals with only minimal computing capability and internet connectivity may make these maps. Hence, they may be an attractive form of communicating information over time for individuals in developing nations.

The drawbacks to this sort of display are as follows:

- Large data sets would not work well in this format if too much needs to be done by hand or offline.
- Although updating might be easy, it might often require manual, rather than automated, manipulation of files.
- Accuracy issues may arise if hand coloring, or other manual data insertion, is done.
- While the 'look' is sufficient to communicate information, it may not appear sleek; some might view it as 'outdated' in appearance.
- There may be a lack of automation in terms of data entry, which may cause difficulties in a number of ways.
- It does not generally take advantage, by its nature, of the collaborative sharing techniques available through cloud computing.

A 3D Approach: Google Earth and Others

Three-dimensional imagery is often quite beautiful and eye-catching. It may draw the reader's eye to the center of the image while it sets other objects or documents to the periphery. A quick use of the mouse by the user may alter all that and change emphasis. Layers can be turned off and on. Animations can be formed, on the fly, by the user. The visual focus is on maps; indirect focus is on 'archive'; when a timeline feature is invoked, there may also be direct focus on a visual timeline that ticks off the years in calendrical fashion (with even, linear spacing in time) with no gaps. Generally, the timeline is displayed as a horizontal timeline. Thus, any text added to the timeline is limited. Google Earth software is free but may require a fairly good computer to make it run efficiently. And it does require an online connection to the internet. It is best suited for use in the developed world and for displays that do not require simultaneous views of the back and the front of the globe.

In Ann Arbor, a series of 3D atlases were formed over a period of years (Arlinghaus et al., 2006–2007). They showed 3D buildings that were cast into a variety of urban contexts, including in a floodplain. Indeed, they even helped to spawn other related projects, according to Naud, such as parts of the City of Ann Arbor Flood Mitigation Plan (2007), which noted especially work done by Paul Lippens and Jerry Hancock.

The merits to this sort of visual display are as follows:

- The creator can bring in data of his or her own and map it and then display it against the onboard data involving terrain, road networks, and other geographic features. This sort of display permits a richness of presentation not available in a simple animated map.
- Mapping using this technology is easy and suggests that as technology continues to improve, mapping is only going to get easier—at least in terms of the mechanics. Of paramount importance then will be a clear conceptual understanding of the sorts of spatial thinking that needs to go on behind the map, in the mind of the map creator.
- Onboard layers of information can be turned off and on by the user. Additional layers may be found on the internet and loaded onto the map. The whole effect can then be saved.
- Individual views may be saved and stacked to form an animated map, offline, in the same way as the simple animated map was created.
- Tours, timelines, and other features are available within the software
- Because the mapping is done online, contemporary computers with good internet connections are required. The software is free; the mechanics to make it run effectively are substantial.
- The software is easy to use and intuitive for many functions. One can also write computer code for more complex features and upload that.

The drawbacks to this sort of visual display are:

- Although the appearance of an image is on a globe, the entire globe cannot be displayed in a single view. One cannot see simultaneously what is happening at one point and at its antipodal point. Thus, creative workarounds are needed to display global data effectively.
- Scholars and casual users in locales with precious connect-time to the internet may have difficulty using this tool.
- It does not generally take advantage, by its nature, of collaborative sharing techniques available through cloud computing.

GEOMATs

The word GEOMAT is an acronym for 'Geographic Events Ordering Maps, Archives, and Timelines'. It was created in the early 2000s, primarily as a teaching tool, by A. Larimore and R. Haug with support from S. Arlinghaus. The visual focus is on a timeline, arranged in vertical fashion, with calendrical spacing of time intervals (no gaps), as a sort of 'march of time'. Because the timeline is vertical in spatial orientation, there is natural room for associated textual commentary, adjacent to the timeline, for readers of languages that are written from left to right or from right to left. (Readers of languages with vertical orientation might wish to employ horizontal timelines for similar reasons.) A map may appear as a root for the timeline. This structure may be presented as an online method on the internet while it is created offline, using a simple web browser that permits html creation in a GUI interface. Or it may be presented offline, in .pdf format created in a table in word-processing software and then communicated as an email attachment. Thus, it is useful in developing world contexts where time online might be slow or limited.

In the sequence of images displayed in Figure 3.1a–c, extracted from a GEOMAT tracking the Wars of the Roses, a small piece of a timeline rooted in an animated Google Earth map, and an associated link, trace the progress of the Wars of the Roses in both space and time as they draw from existing archival materials. Figure 3.1d shows a QR code to link to the full file.

The merits to this sort of visual display are as follows:

- The vertical timeline permits the readable inclusion of substantial amounts of text.
- The animated map provides a compact view of the sequencing of spatial relations over the entire war.
- The nested timelines offer a means to communicate large amounts of temporal data (as they did with the geological timeline).
- Individuals in developing nations can create these using only a simple GUI browser that can be downloaded free and then used offline to create as intricate a display as they wish. Precious connect-time to the internet is minimized in a number of locales.

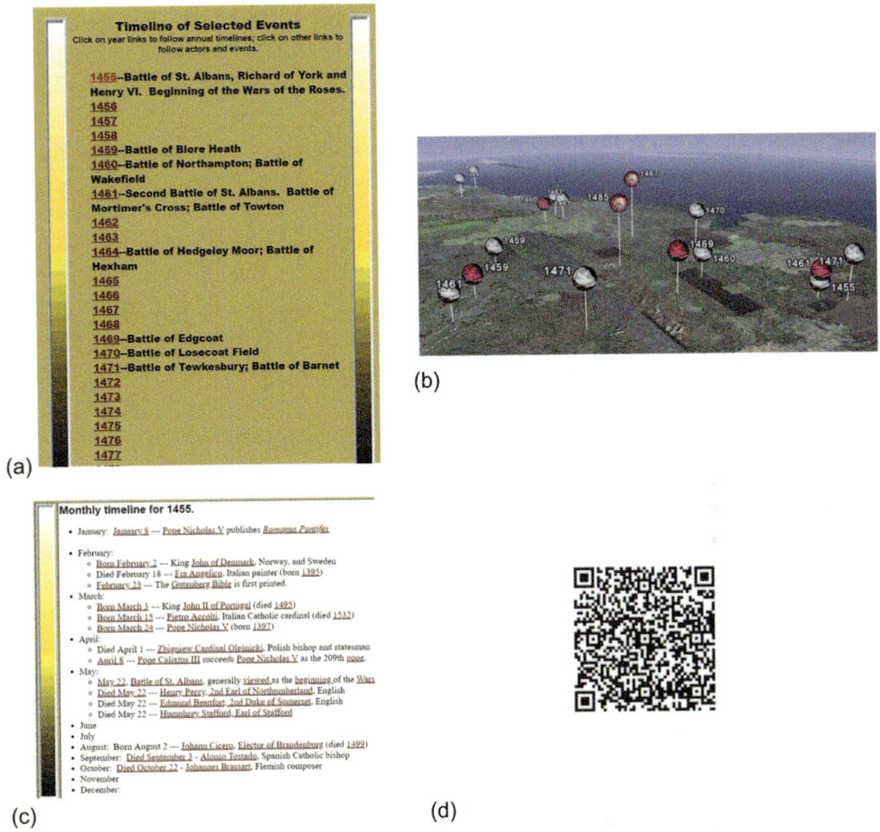

Figure 3.1 *GEOMAT of the Wars of the Roses. (a) Timeline associated with the Wars of the Roses. (b) Animated 3D map in which the timeline is rooted. (c) Nested timeline associated with the primary timeline, accessible through a link on the appropriate year. (d) QR code link to the entire GEOMAT: http://www-personal.umich.edu/~sarhaus/MapsAndTimelines/index.html. (Source: IMaGe, ©2008. Used with permission.)*

- Using a browser to create the material has the intellectually elegant advantage of using the tool for creation also as the tool for display.
- The software is easy to use and intuitive, requiring no special expertise beyond standard computer literacy.

The drawbacks to this sort of visual display are as follows:

- The display does not take advantage of the most contemporary software. As with the simple animap, much is done by hand within easy-to-use software.

- Because timelines are calendrical and vertical, they may cover many pages for studies covering long periods of time with long time gaps between known events. Creative management of this issue might involve nesting timelines.
- It does not generally take advantage, by its nature, of the collaborative sharing techniques available through cloud computing.

Story Maps

Like the GEOMAT, the story map can simultaneously present all three elements—maps, archives, and timelines—with each element evident visually. Unlike the GEOMAT, it generally relies on having a good, high-speed internet connection. It is convenient for users with such capabilities to be able to create a story map online, and most important perhaps, it does take advantage of the power of collaborative cloud computing and web mapping. It is not, however, useful in environments with very limited internet availability.

One simple story map, which tells the story of the diffusion of influenza across the United States, is available annually from the Centers for Disease Control (Figure 3.2a,b). The horizontal timeline enables the user to change the map to track changes over time. Thus, in Figure 3.2a, we see the spread by state in the middle of the flu season of 2017. In Figure 3.2b, we see it at the end of that season in the spring of 2018. Our minds interpolate the spread when given only two static images. Figure 3.2c provides a QR code and a link to the full map, which tells a much richer and more detailed story.

This particular story map tells the full story of a very well-defined, and clearly circumscribed, single style of event. The Esri story map site is filled with story maps that tell much broader and more complex stories (Esri, 2018a). One contemporary story map on that site is devoted to "Scientists Sharing Research" (Esri, 2018b). The associated Esri website offers people the opportunity to use this form of Esri software for free (much of their software is for pay). Kerski's chapters clearly reflect on the importance of story maps within the broad context of web mapping and the substantial changes that are emerging surrounding contemporary shifts in the handling of spatial data and analysis. We explore several of these broader story maps in this book.

The merits to the story map visual display are as follows:

- It has the capability to include all the merits noted in the three previous forms of display.
- It has the ability to create displays online.
- It is updated in a continuing manner, by a dedicated (Esri) staff, to include capabilities of the most recent software, spatial data, and related technological advances.
- It takes advantage, by its nature, of the collaborative sharing techniques available through cloud computing and web mapping.

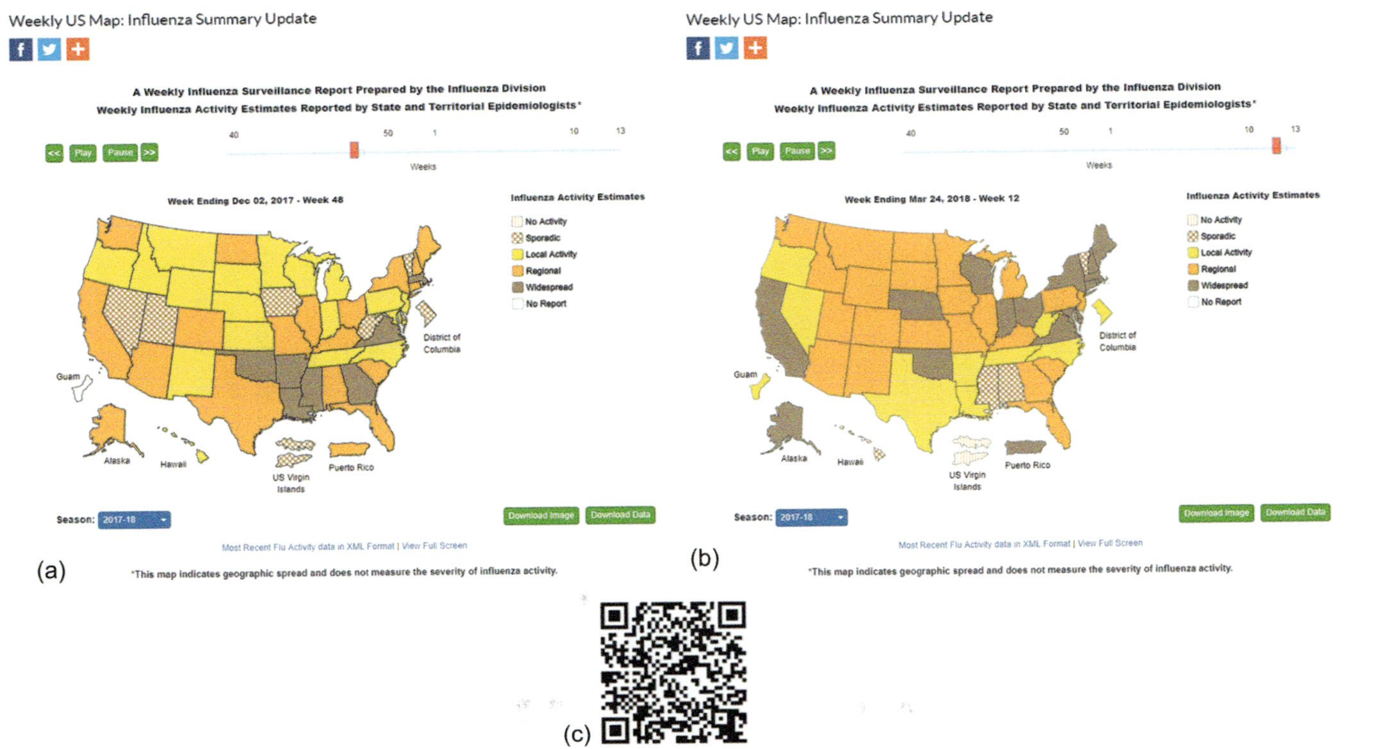

Figure 3.2 *(a) and (b) show the diffusion of influenza, by state in the United States, in early December 2017 and late March 2018 (respectively). (c) is a QR code linking to the full CDC site: https://www.cdc.gov/flu/weekly/usmap.htm. (Source: CDC, 2018.)*

The drawbacks to this sort of visual display are as follows:

- There is a need for contemporary computing equipment and internet access.
- It may not be readily available for individuals in developing nations.

Observations

In 2013, Arlinghaus and Kerski published the first book, *Spatial Mathematics: Theory and Practice through Mapping*, in this CRC Series devoted to Spatial Science. They remain hopeful that their nudgings in that book—concerning the enduring prospect of breaking down the barriers between disciplines—are moving forward in various, perhaps unforeseen, directions. Kerski observes, from his incessant travel as a GIS educator for Esri, that some serious headway in that regard is being made in some universities' schools of business and in their departments of sociology, engineering, history, and a few others, such as conflict and peace studies, which may have been unlikely to look at geotechnologies or spatial thinking in the past but now are starting to ask, and to delve into, spatial questions. Naud has been using it in municipal government applications in equity and sustainability studies. That change in understanding is very encouraging and is responsive to our general goal of fostering spatial thinking in creative ways.

Indeed, with all the wonderful technological advances that we see, far too often the context and the big picture get lost. We learn one version of spreadsheet software, then an update comes along, then another, and then another. Those of us with a broad liberal arts background or a specialized scientific background, coupled with an open mind, extract general principles in the structure of the software and adjust, fairly easily, to the new updates. The context matters. Indeed, such abstract context, characterized as 'spatial mathematics', was also a focus, along with 'spatial thinking', of the first book in this series (Arlinghaus and Kerski, 2013). Here we take a similar approach but with a different real-world focus: the emphasis now is on application (in environmental contexts) rather than on theory (in mathematical contexts).

We believe this book will make a contribution to the continued outreach to a wide range of disciplines, with the message "Spatial thinking matters" and "You don't have to be a GIS expert to use some of these tools." Thus, 'mathematical/theoretical context' transforms into 'environmental/application-oriented context' with the addition to the team of Matthew Naud, environmental coordinator of the City of Ann Arbor, and Ann Larimore, cultural geographer. The team thus extends naturally into urban and historical environmental contexts, always with the goal of encouraging systematic and logical thinking about the spatial matters in the world.

References

Arlinghaus, S. 2008. The Wars of the Roses, 1455–1487: Animation and interaction as space/time transformations. Retrieved from http://www-personal.umich.edu/~sarhaus/MapsAndTimelines.

Arlinghaus, S. et al. 2006–2007. *3D Atlas of Ann Arbor*, 1st, 2nd, and 3rd Editions. http://www.imagenet.org.

Arlinghaus, S., and Kerski, J. 2013. *Spatial Mathematics: Theory and Practice through Mapping*. Boca Raton: CRC Press.

City of Ann Arbor. 2007. Flood mitigation plan. Retrieved from https://www.a2gov.org/departments/systems-planning/planning-areas/water-resources/floodplains/Documents/FloodplanMitigationPlan_Mar07.pdf.

Esri. 2018. Story maps. Retrieved from https://storymaps.arcgis.com/en.

Esri. 2018. Scientists sharing research. Retrieved from https://science.maps.arcgis.com/apps/MapSeries/index.html?appid=247cd4da5af140eb9e9055e13ca5b94f.

FluView. 2018. Centers for Disease Control and Prevention. Retrieved from https://www.cdc.gov/flu/weekly/usmap.htm.

NASA. 2013. Video: Global temperature variation. Retrieved from https://climate.nasa.gov/climate_resources/101/video-global-temperature-variation.

II

Animaps, 1990s

The section begins with an animated map, one forerunner of a display style of integrating space and time that was quite exciting when first introduced in 1998. Today, while not avant-garde, it may remain a useful display that is easy to make offline using simple software. The next chapter illustrates a display that derived from the ideas in the first chapter, suggesting directions of evolution. The section concludes by intertwining concepts with practice from the first two chapters.

4

Animaps: Varroa Honeybee Mite

Diana Sammataro, Sandra L. Arlinghaus, and John D. Nystuen

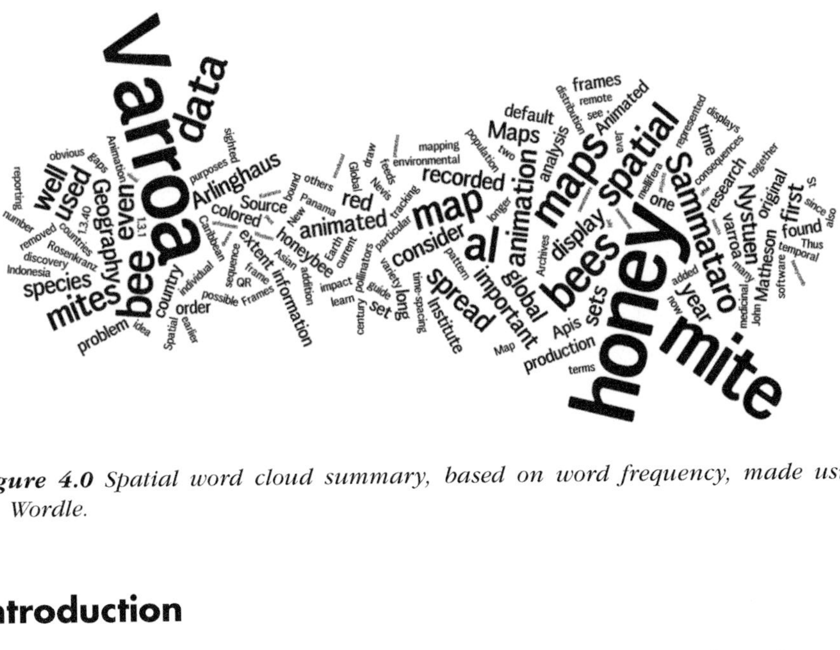

Figure 4.0 *Spatial word cloud summary, based on word frequency, made using Wordle.*

Introduction

Varroa (Acari: Varroidae) is a parasitic mite that poses a serious threat to the European honeybee (*Apis mellifera*) (Sammataro and Avatible, 1973, 1st ed.; 2018, 5th ed.). This mite, related to ticks, feeds on adult and larval honeybees, obtaining nourishment from their blood and the fat bodies carried in the bees' bloodstream. This mite not only feeds on bees and bee larvae but carries viral diseases and promotes stress to the hard-working bees (Sammataro et al., 2000). We have been mapping the spread of this blight for quite a while, focusing on the importance of animation as a tool to draw together space and time. Understanding spatial patterns helps to tighten the focus on intervention. There are obvious consequences associated with the possible extinction of honeybees: honey has long been an important agricultural crop (Ellis and Munn, 2005; Matheson, 1996). In addition, honeybees are important pollinators of one-third of our crops, including fruits and vegetables, and are used in seed production in a variety of other plants (Free, 1993; McGregor, 1976).

There are substantial economic implications to the possible demise of honeybee pollinators as well as to the production of honey, long used as a natural sweetener (a healthy alternative to processed sweeteners) and for medicinal purposes. The production of beeswax from the honeycomb is even more valuable than honey, as it is a primary foundation for cosmetics, for candles, and for art projects and even has medicinal uses. All of these hive products have been important since beekeeping was first recorded; wax and propolis (bee-collected plant resins) were vital to preserving Egyptian mummies. Beyond the obvious, when an established species is removed from an ecosystem it is simple logic that the impact of such removal will have long-range, and perhaps unforeseen and unintended, consequences.

The Varroa problem began in Asia in the early 20th century (De Jong et al., 1984; Rosenkranz et al., 2010). Today, Varroa is found worldwide, with some exceptions (Bradbear, 1988; Matheson, 1996), such as Australia. Erroneous classification of the mite has clouded some of the reporting of information. First identified as *Varroa jacobsoni* on the Asian honeybee (*Apis cerana*), molecular analysis has now separated out four different Varroa species. We refer here, for purposes of mapping, to the mite simply as Varroa, and in general terms it represents the new *Varroa destructor* (Anderson and Trueman, 2000) that jumped from *A. cerana* onto *A. mellifera*. Careful analysis of the problem as a whole, beyond the tracking aspects, must consider the taxonomic problems as well (Rosenkranz et al., 2010; Navajas et al., 2010).

As late as 2000, Varroa was discovered in New Zealand (Matheson, 2015), Panama (Calderon et al., 2000) and St. Kitts & Nevis in the Caribbean. It has also been found in the Caribbean islands of Grenada in 1994, Trinidad in 1996 (Hallim, 2000), Cuba in 1996, Dominica in 1998, St. Lucia in 1999, and Tobago and Nevis in 2000. It has apparently also been reported in Haiti. On July 6, 2000, Varroa was first detected in Panama.

The discovery of Varroa mites in the Eastern Rift Valley in eastern Kenya (Frazier et al., 2010), the homeland of the honeybee species as well as a diverse population of wild (often unusual) animals, is particularly alarming because bees and honey are an integral part of subsistence-level farming, where honey is an important source of income. The discovery of mites somewhat earlier (Kunimoto, 2007) in the tropical paradise of the remote Hawaiian Islands, will have a huge eventual impact since many breeders raise queen honeybees there. The spread of these mites can be directly attributed to the movement of bee colonies by beekeepers and as well as from some hitchhiking bee swarms on ships.

The plight of the honeybee is a global issue; we capture it here in an animated visual format that is easier to digest than straight text reporting. Thus, students and researchers can learn from the past and consider appropriate environmental interventions for the future.

On the horizon is another mite, *Tropilaelaps sp.*, that is beginning to be of concern (see http://entnemdept.ufl.edu/creatures/MISC/BEES/Tropilaelaps.htm). And the tracheal mite, (*Acarapis woodi*) once the only introduced mite in the New World, is now generally overlooked in light of the other mites that are on the honeybees. They are still to be found, however, and should not be forgotten.

A Tale of Two Maps: The Evidence

The pair of maps shown in Figures 4.1 and 4.2 display the global extent of Varroa mite presence by country. The first recorded sightings, in 1904, were on the island of Java in Indonesia (Figure 4.1). That sighting is represented in red in the inset map. As a consequence of that single sighting, the entire country is colored red; that does not mean that the mite was everywhere, just that there was at least one sighting. The same strategy remains as other countries enter the picture.

Over time, the mite has spread to threaten the global honeybee distribution. Figure 4.2 displays the extent of the expansion by 2011. The countries added in 2011 are colored in red, while those entering the picture in earlier times are colored a shade of muted red (pink). Clearly, by 2011, the spatial extent had penetrated even remote corners of the Earth.

What happened in between the spatial distributions recorded in these two maps? There are a number of issues to consider. The basic analysis is a bounded one: the temporal bound is from 1904 to the present, a bit more than one century; the spatial bound is the non-polar land surface of the Earth. With the bounds in mind, we consider ways to look at the spatial and temporal components of the problem, in order to draw them together into a package designed to shed insight on the process to learn from this particular environmental disaster as a guide to avoiding others.

Animated Map

Existing records provide information as to where Varroa mites were sighted, by country, and when they were first sighted, by year. Some of the data sets are more accurate than others, but location by country and by year only is

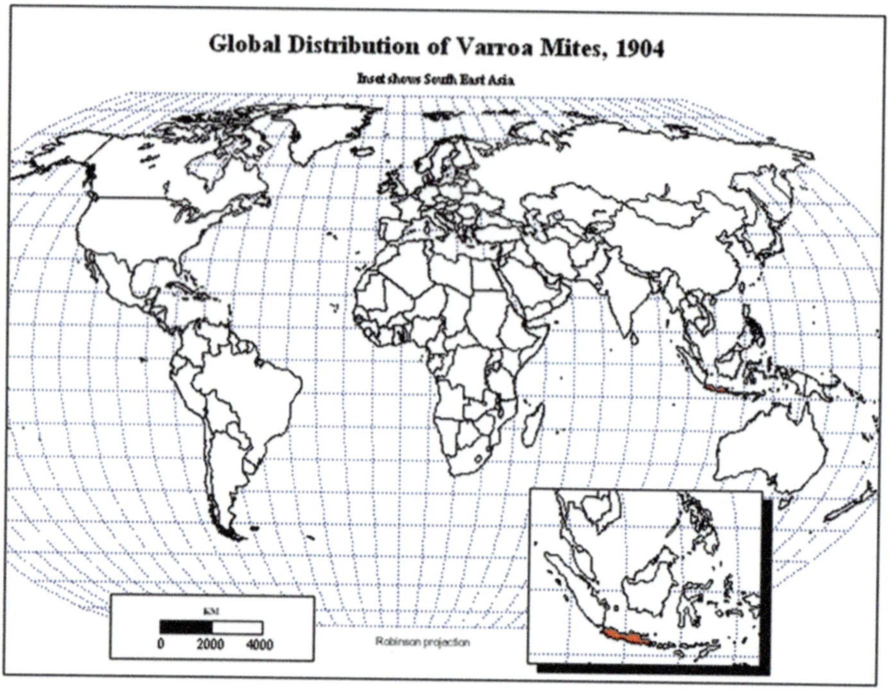

Figure 4.1 *Varroa mites recorded in Java (Indonesia). (Source: IMaGe, ©1998, Solstice, http://hdl.handle.net/2027.42/58289. Used with permission.)*

itself a rough, but useful, spatial display. It is a straightforward matter to map the data sets and then to animate them as a .gif. We used Adobe software to execute this and colored the countries in the same scheme as in Figures 4.1 and 4.2. In addition to the two preceding maps, the following maps show the diffusion of the mite by year (Figure 4.3.1–4.3.4).

Arlinghaus created the original mapped set, up to 1998, in order to support her idea to make an animated map to clearly display the global spread of the Varroa mite and other phenomena across the Earth. She used data from Sammataro for one of these 'animaps'. With the completed map in hand, John Nystuen contributed the idea of distinguishing the current fringe at any time point using a unique color (red), in order to emphasize the particular year and track the spread.

In terms of timing, the data set had information for some, but not all, of the years in the course of the original study. There were gaps in the data. Animation software, by default, set the time-spacing between default frames at a constant value. Frames came in and out of the animation at an even rate and many maps were originally used to represent long periods

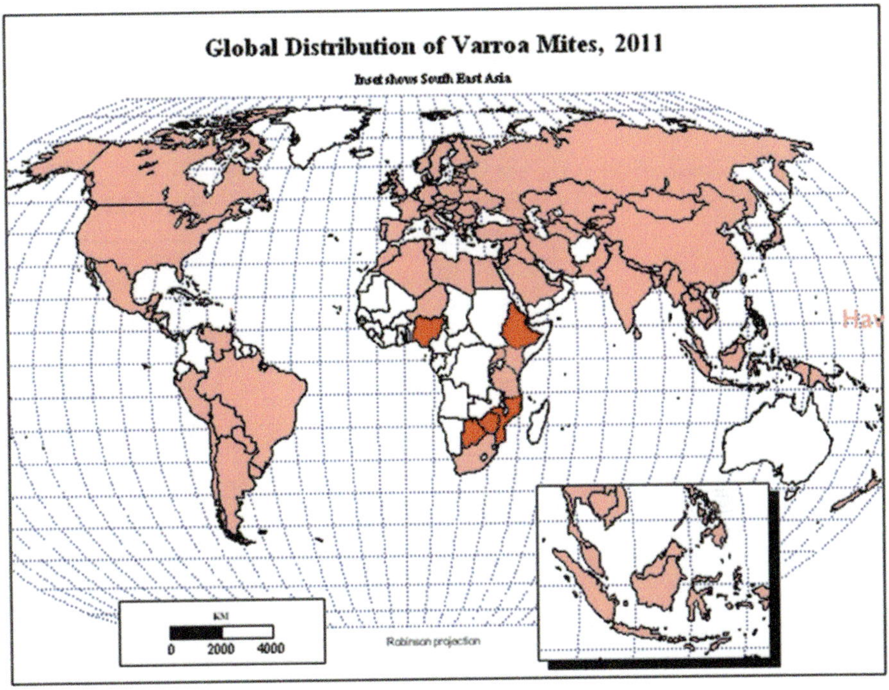

***Figure 4.2** Varroa mite distribution as of 2011. (Source: IMaGe, ©2012 http://hdl.handle.net/2027.42/94573. Used with permission.)*

of inaction. Nystuen suggested overriding the default so that longer gaps in the data sets would be represented by longer time-spacing between successive animation frames, thereby reducing the number of individual maps in the animation (Arlinghaus et al., 1998). Thus, the sequence of individual frames, when removed from the animation, displays no maps from 1912 to 1947. A frame for 1948, albeit the same as the 1912 frame, was inserted for emphasis.

Arlinghaus has updated the original animated map on a regular basis as Sammataro has acquired more data sets. These updates have been published regularly in the journal *Solstice: An Electronic Journal of Geography and Mathematics* (Sammataro, 2001; 2006; 2007; 2011; 2012; Sammatro and Arlinghaus, 2010). The QR code in Figure 4.4 links to the current state of the animated Varroa map.

While the map sequence does display a clear spatial pattern, it does not offer any substantive research content. It serves as a base from which to organize research and as a guide to evoke questions that might lead to additional research.

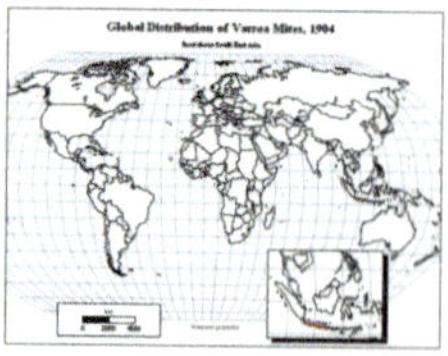

Figure 4.3.1: 1904

Figure 4.3.2: 1912

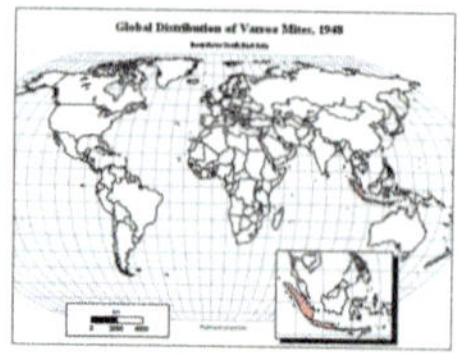

Figure 4.3.3: 1948

Figure 4.3.4: 1949

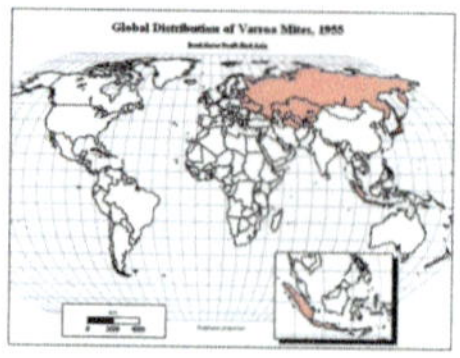

Figure 4.3.5: 1955

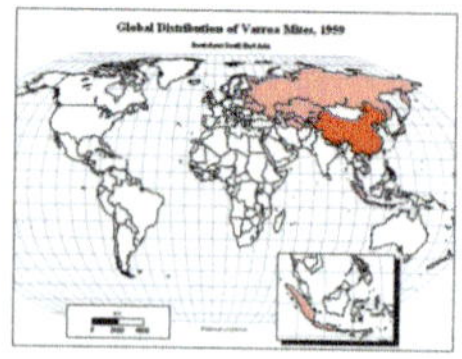

Figure 4.3.6: 1959

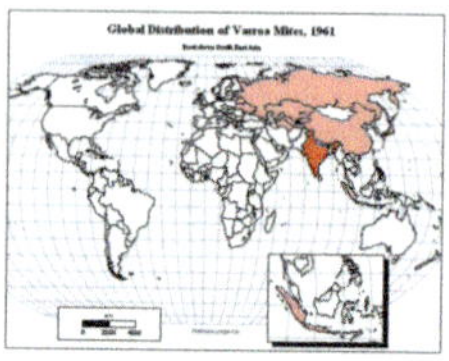

Figure 4.3.7: 1961

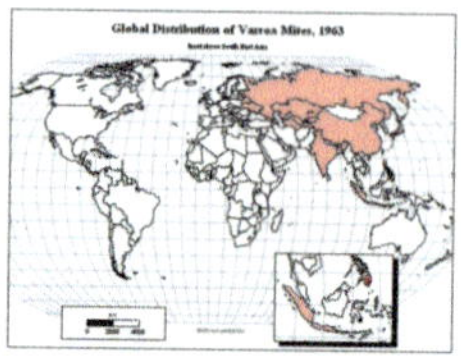

Figure 4.3.8: 1963

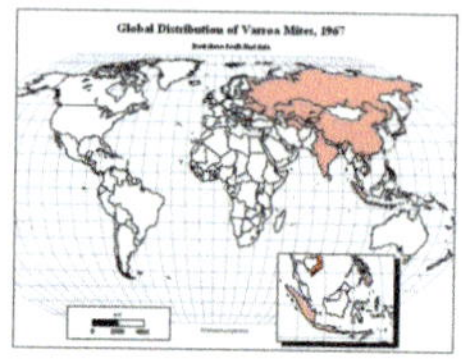

Figure 4.3.9: 1967

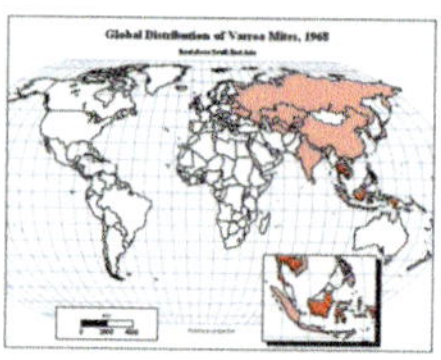

Figure 4.3.10: 1968

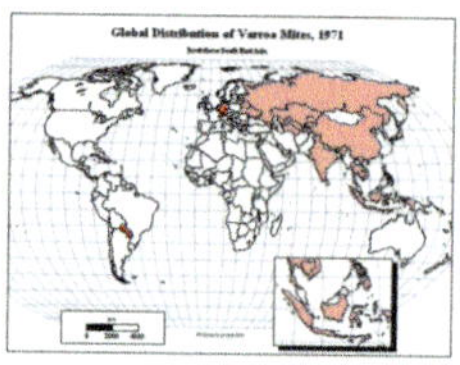

Figure 4.3.11: 1971

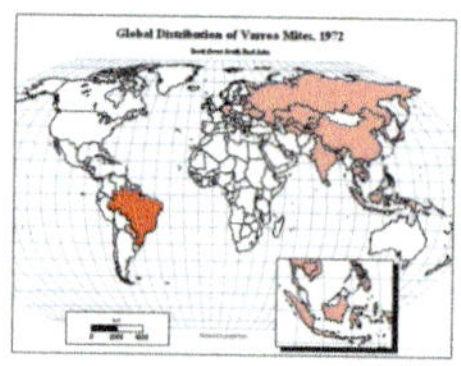

Figure 4.3.12: 1972

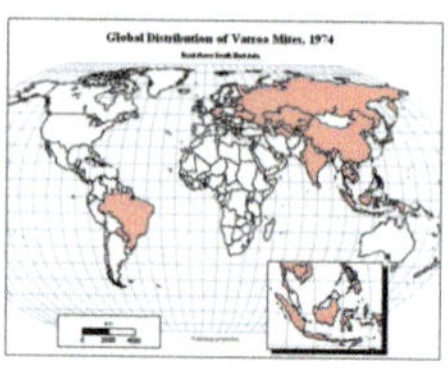

Figure 4.3.13: 1974

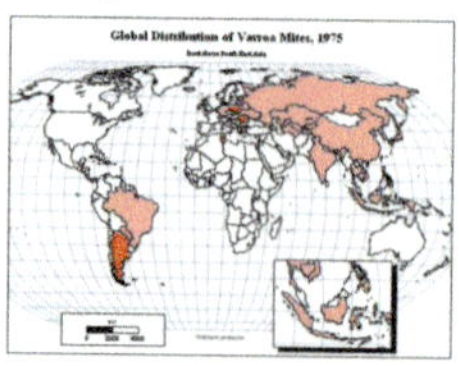

Figure 4.3.14: 1975

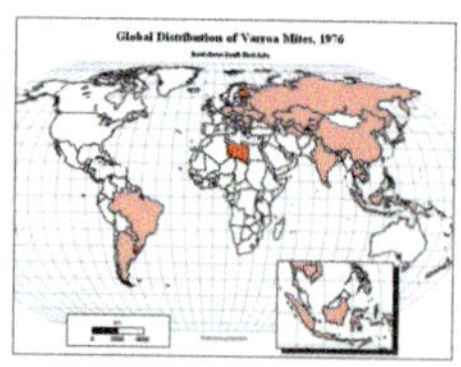

Figure 4.3.15: 1976

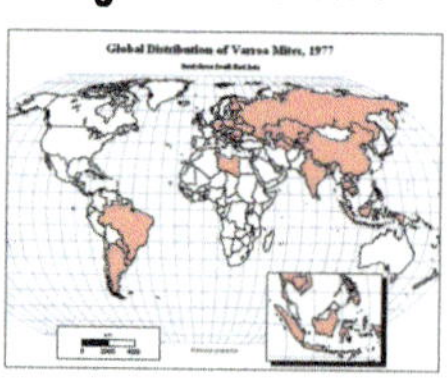

Figure 4.3.16: 1977

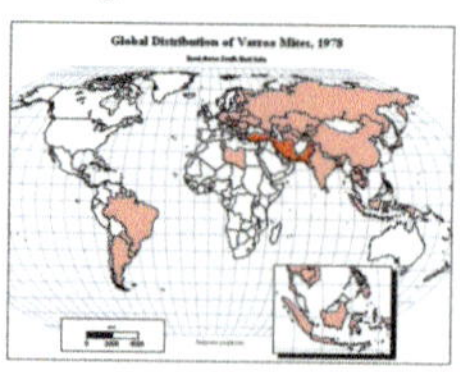

Figure 4.3.17: 1978

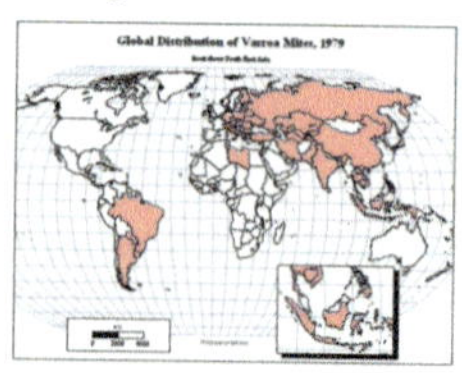

Figure 4.3.18: 1979

Figure 4.3 *(Continued)*

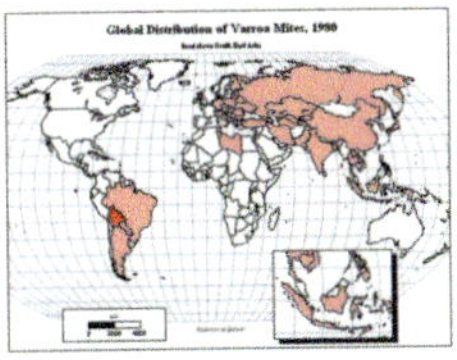
Figure 4.3.19: 1980

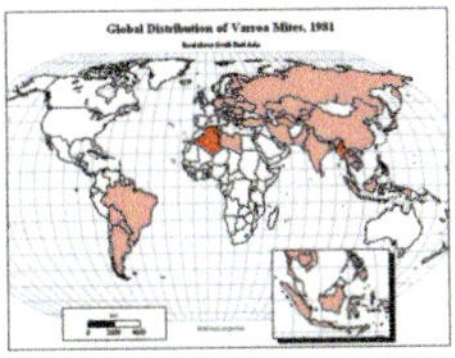
Figure 4.3.20: 1981

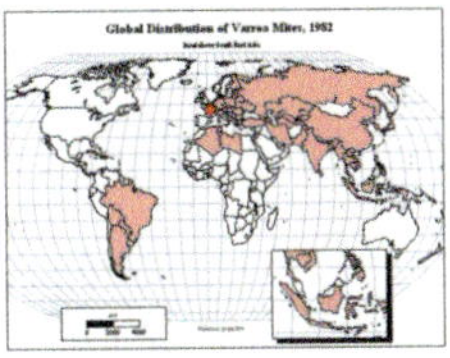
Figure 4.3.21: 1982

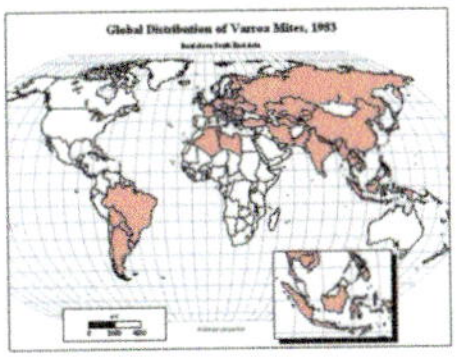
Figure 4.3.22: 1983

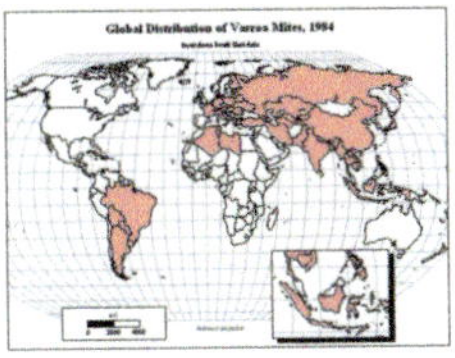
Figure 4.3.23: 1984

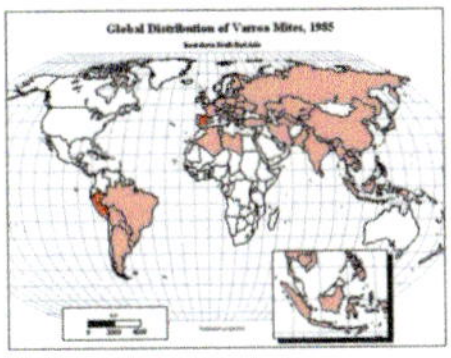
Figure 4.3.24: 1985

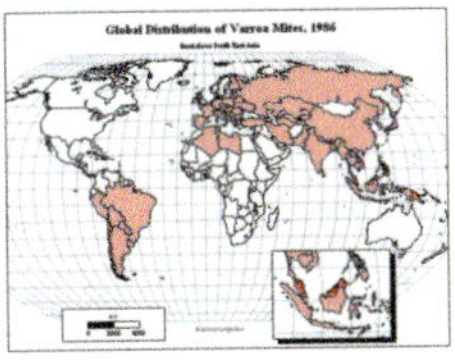
Figure 4.3.25: 1986

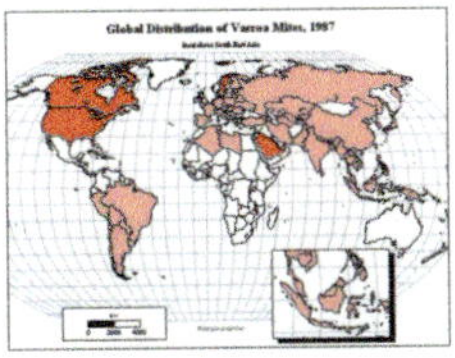
Figure 4.3.26: 1987

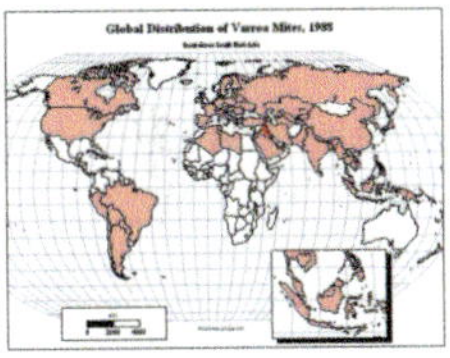
Figure 4.3.27: 1988

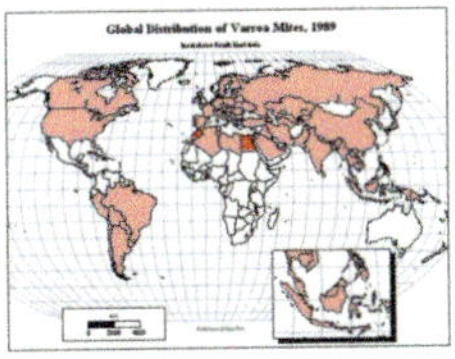
Figure 4.3.28: 1989

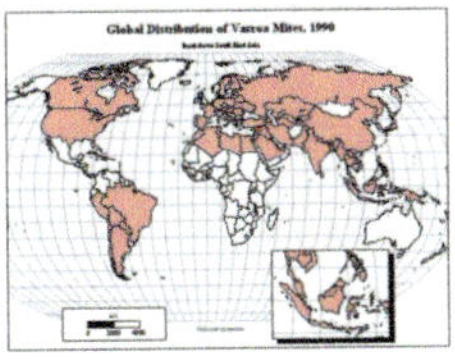
Figure 4.3.29: 1990

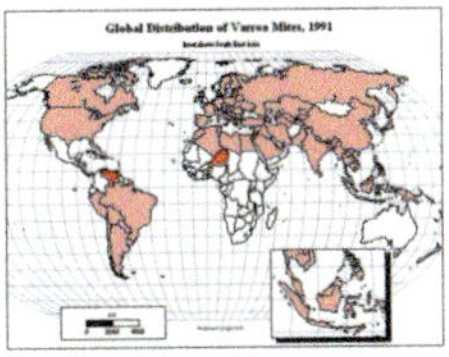
Figure 4.3.30: 1991

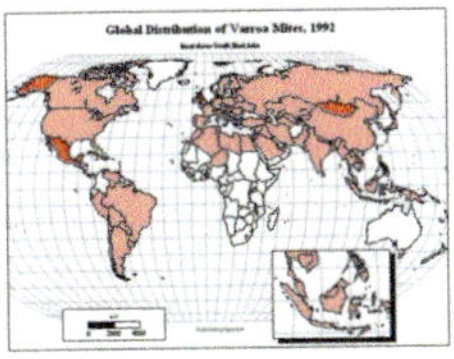
Figure 4.3.31: 1992

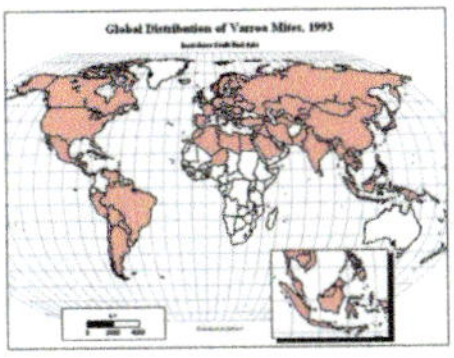
Figure 4.3.32: 1993

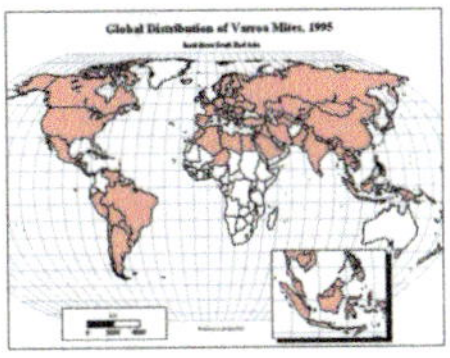
Figure 4.3.33: 1995

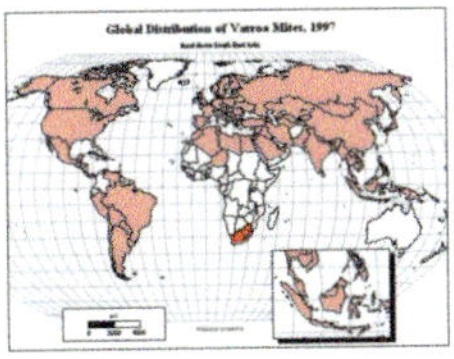
Figure 4.3.34: 1997

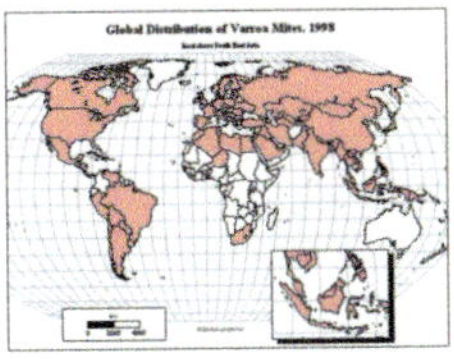
Figure 4.3.35: 1998

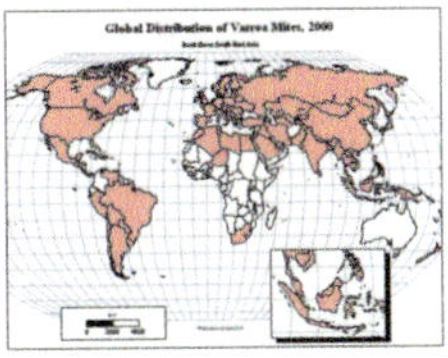
Figure 4.3.36: 2000

Figure 4.3 *(Continued)*

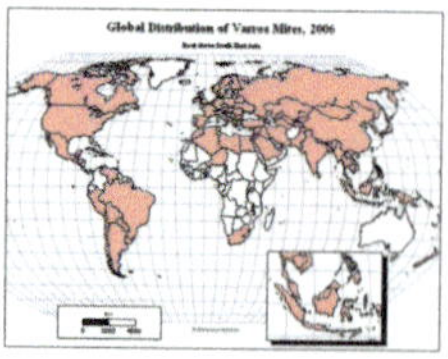

Figure 4.3.37: 2006

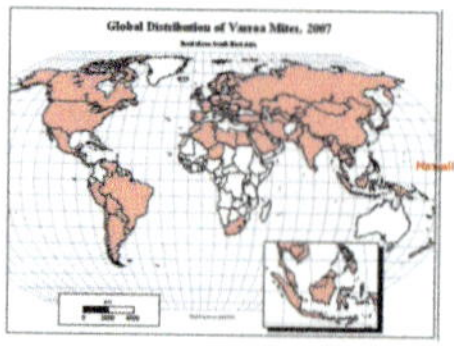

Figure 4.3.38: 2007

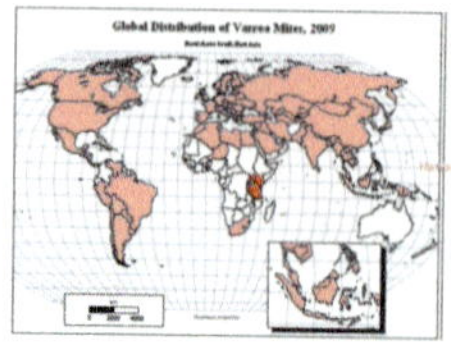

Figure 4.3.39: 2009

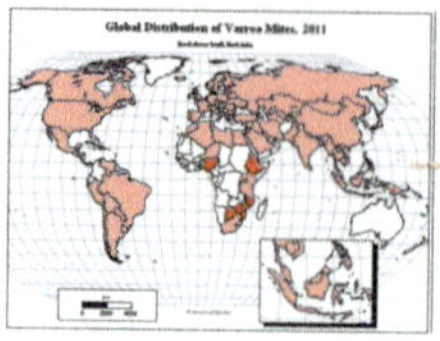

Figure 4.3.40: 2011

Figure 4.3 *(Continued) Animation frames to fill the gap between Figures 4.1 and 4.2. (Source: IMaGe, ©1998–2012. Used with permission.)*

Figure 4.4 *QR Code to link to animated map. (Source: (http://www.mylovedone.com/image/solstice/win12/varroa2012b.html).)*

References

Anderson, D. L., and J. W. H. Trueman. 2000. *Varroa jacobsoni* (Acari: Varroidae) is more than one species. *Experimental & Applied Acarology* 24: 165–189.

Arlinghaus, S. L., W. D. Drake, and J. D. Nystuen, with data and other input from A. Laug, K. S. Oswalt, and D. Sammataro. 1998. Animaps. *Solstice: An Electronic Journal of Geography and Mathematics* 9(1).

Bradbear, N. 1988. World distribution of major honeybee diseases and pests. *Bee World* 69(1).

Calderon, R. A., A. Ortiz, B. Aparisio, and M. T. Ruiz. 2000. Varroa in Panama: Detection, spread and prospects. *Bee World* 81(3): 126–128.

De Jong, D., L. S. Goncalves, and R. A. Morse. 1984. Dependence of climate on the virulence of *Varroa jacobsoni*. *Bee World* 65: 117–121.

Ellis, J. D., and P. A. Munn. 2005. The worldwide health status of honey bees. *Bee World* 86(4): 88–101. http://www.ibra.org.uk.

Frazier, M., E. Muli, T. Conklin, D. Schmehl, B. Torto, J. Frazier, J. Tumlinson, J. D. Evans, and S. Raina. 2010. A scientific note on *Varroa destructor* found in East Africa: Threat or opportunity? *Apidologie* 41: 463–465. Online: RA/DIB-AGIB/EDP Sciences, 2009. http://www.apidologie.org. doi:10.1051/apido/2009073.

Free, J. D. 1993. *Insect Pollination of Crops*, 2nd Edition. London: Academic Press. 684 p.

Hallim, M. K. I. 2000. Pests and diseases of honeybees in Trinidad and Tobago in the year 2000 and recommendations to reduce their spread in the Caribbean. Paper presented to the Second Caribbean Beekeeping Congress, Nevis, August 14–18, 2000. Ministry of Agriculture Fisheries and Food, UK (1996). Varroosis: A parasitic infestation of honeybees.

Kunimoto, S. L. 2007. Varroa mite survey, State of Hawaii, https://hdoa.hawaii.gov/pi/files/2013/01/npa07-01-Varroa.pdf.

Matheson, A. 1996. World bee health update 1996. *Bee World* 77: 45–51.

Matheson, A. 2015. World bee health update 2015. *Bee World* 74(4): 176–212.

McGregor, S. E. 1976. Insect pollination of cultivated crop plants. *Agriculture Handbook* No. 496. Washington, DC: ARS-USDA. http://www.ars.usda.gov/SP2UserFiles/Place/53420300/OnlinePollinationHandbook.pdf; http://www.sciencedaily.com/releases/2007/05/070510114621.htm.

Navajas, M., D. L. Anderson, L. I. De Guzman, Z. Y. Huang, J. Clement, T. Zhou, and Y. Le Conte. 2010. New Asian types of *Varroa destructor*: A potential new threat for world apiculture. *Apidologie* 41(2): 181–193.

Rosenkranz, P., P. Aumeier, B. Ziegelmann. 2010. Biology and control of *Varroa destructor*. *Journal of Invertebrate Pathology* 103(suppl. 1): S96–S119.

Sammataro, D. 2001. Update from Diana Sammataro: Varroa mite animated map. *Solstice: An Electronic Journal of Geography and Mathematics* 12(1).

Sammataro, D. 2006. Update on the Varroa mite map. *Solstice: An Electronic Journal of Geography and Mathematics* 17(2).

Sammataro, D. 2007. Update on the Varroa mite map [with editorial commentary]. *Solstice: An Electronic Journal of Geography and Mathematics* 18(1).

Sammataro, D. 2011. Varroa mite project. Note in *Solstice: An Electronic Journal of Geography and Mathematics* 22(2).

Sammataro, D. 2012. Update on Varroa mite spread. *Solstice: An Electronic Journal of Geography and Mathematics* 22(2).

Sammataro, D., and S. L. Arlinghaus. 2010. The quest to save honey: Tracking bee pests using mobile technology. *Solstice: An Electronic Journal of Geography and Mathematics* 21(2).

Sammataro, D., and A. Avitabile. 1973. (1st Edition; 5th Edition now underway 2018). *The Beekeepers Handbook*. Ithaca, NY: Cornstock Publishing Associates, Cornell University Press.

Sammataro, D., U. Gerson, and G. Needham. 2000. Parasitic mite of honey bees: Life history, implications and impact. *Annual Review of Entomology* 45: 519–548.

5

Animap Timelines: The Space-Time Pattern of Cutaneous Leishmaniasis, Syria 1990–1997

Salma Haidar, Mark L. Wilson, and Sandra L. Arlinghaus

Figure 5.0 Spatial word cloud summary, based on word frequency, made using Wordle.

Leishmaniasis represents a group of tropical diseases caused by various species of Leishmania parasites that are transmitted by the bite of sandflies. An estimated 12 million people globally are affected, with approximately 1.5–2.0 million new cases annually. It is generally an ulcerative disease that may cause lesions on the outside of the body or internally, along with the consequent effects of that ulceration. Cutaneous leishmaniasis occurs principally in certain developing nations of the Middle East, Africa, Asia, and South America, typically among the world's poorest peoples. Risk factors for disease include poverty, malnutrition, deforestation, lack of sanitation, and urbanization. Cutaneous leishmaniasis is a disease with both demographic and environmental associations (WHO, 2013; Haidar, 2002). The World Health Organization has classified it as a neglected tropical disease.

Animation Rationale

Cartographic evidence can often be used to find patterns in large sets of data that are widely scattered in time and space. Thus, when co-author Haidar considered spreadsheets with many thousands of entries related to the presence of cutaneous leishmaniasis in Syria, it seemed useful to map the data in her quest to look for patterns in incidence of the disease. She wished to view the data by Syrian province (Figure 5.1) over a period of 8 years, on a monthly basis. In that way she hoped to be able to see, at a glance, variations in incidence from north to south in a seasonal framework. The animated map offered one approach to that task (Arlinghaus, Haidar, and Wilson, 2002).

Leishmaniasis Data for Provinces of Syria: 1990 by Month

To create a sequence of maps, we employed a sequence of steps, using contemporary geographic information system (GIS) software to gain benefit from the analytic capability of the software. Monthly thematic maps were created for each year (a total of 8 × 12 = 96 maps). Figures 5.2a–5.2l show the set of 12 maps for the year 1990, arranged by month. The maps were colored

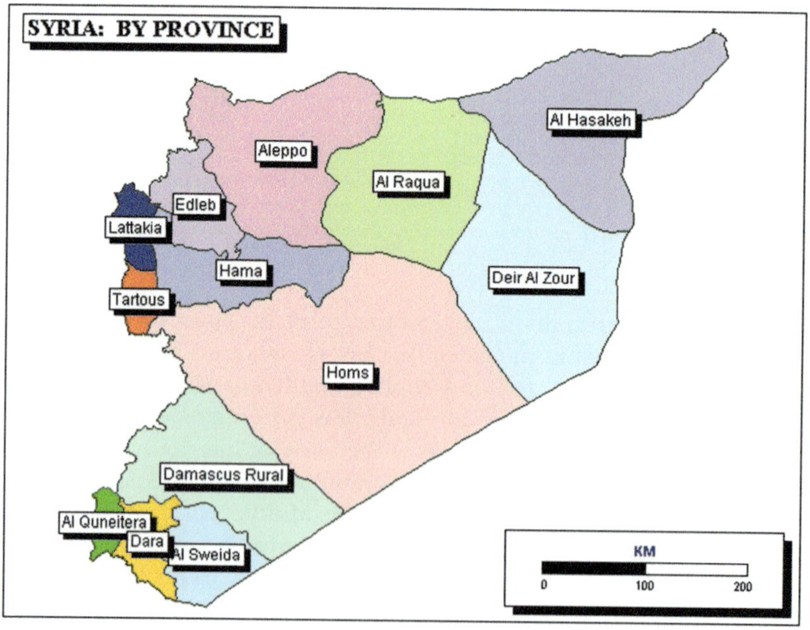

Figure 5.1 *Provinces of Syria. (Source: IMaGe, ©2002. http://hdl.handle.net/2027.42/58243. Used with permission.)*

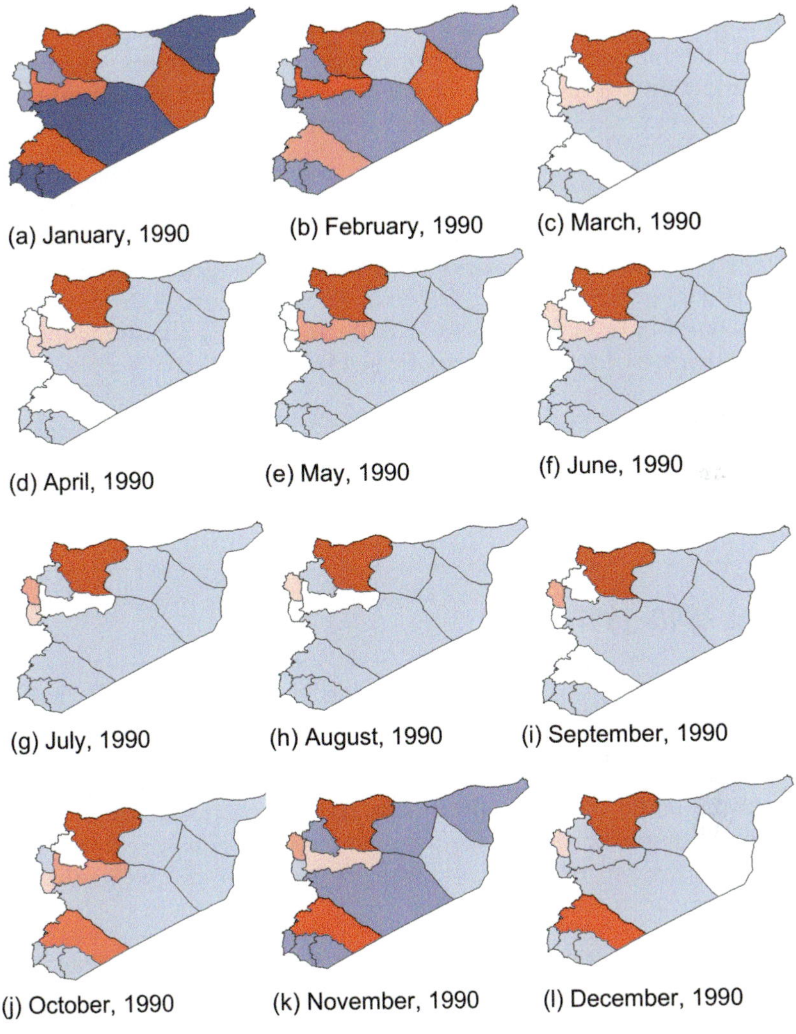

Figure 5.2 *1990: Leishmaniasis by Syrian province by month. Red indicates incidence above the mean (in white); blue indicates incidence below the mean. The deeper the color, the further from the mean. (Source: IMaGe, ©2002. http://hdl.handle.net/2027.42/58243. Used with permission.)*

as thematic maps using standard deviation in the data by 0.25 SD per color interval. Provinces were colored red if the number of cases of the disease for that month was above the mean, blue if below the mean, and white if near the mean. The deeper the color, the further the number of cases was from each monthly mean.

Animation of Leishmaniasis Data by Province: 1990–1997

One can see patterns by looking at the monthly static images over the period of a year. In this sequence for 1990, the province of Aleppo seems to be a center for activity, independent of the time of the year. One might imagine intervention centered in Aleppo to be one strategy.

To get a different view of the sequence, we animated it, with the advantage of having done some analysis (in relation to the mean) in advance of animation. While patterns can be retained in the mind over only 12 images, other patterns may pop out in an animation when the 12 images are compressed into one frame. Figure 5.3 contains a QR code to the base article from which some of the material here is derived. In it, one can watch a simultaneous, month-by-month display of the animations from 1990 to 1997.

- Non-persistent link: http://www-personal.umich.edu/%7Ecopyrght/image/solstice/sum02/animapssyria.html
- Persistent link (scroll down to *Solstice* 13(1)): https://deepblue.lib.umich.edu/handle/2027.42/58219

July Data: 1990–1997

Animation by year offered substantial compression of data into eight yearly files, each with 12 images. While one can perhaps examine patterns over the course of a single year of static images, it is not likely that a reader could hold 96 images from all 8 years in his/her mind and extract all 'July' images. Comparisons made over a number of years at a particular time of the year reveal among-year, rather than within-year, patterns. Figure 5.4a–h show eight images for July, for example, illustrating change over time cutting through the 12 × 8 data sheaf in another direction.

Again, one can more easily retain a sequence of eight images such as these. Additional animations in this direction, through annual sequences, may also be found in the link in Figure 5.3. Again, it appears that disease incidence

Figure 5.3 *QR code linking to animations of Leishmaniasis prevalence, by month, in Syria.*

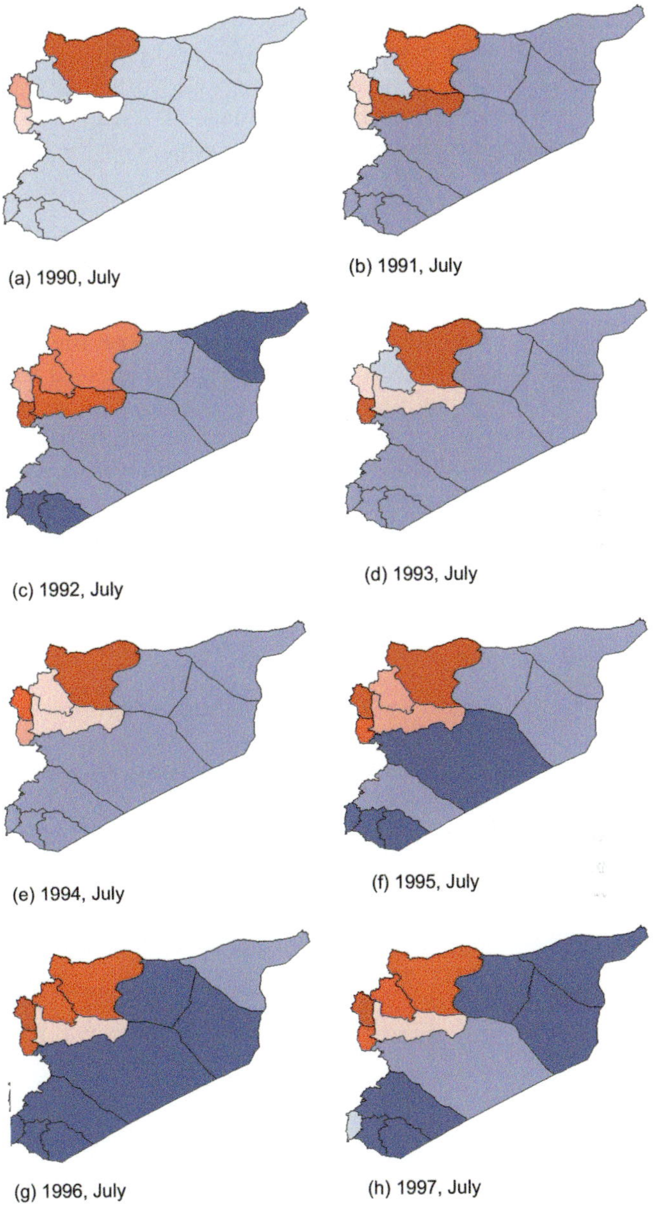

Figure 5.4 *Incidence of leishmaniasis in Syria, by province, over a period of 8 years in July. (Source: IMaGe, ©2002. http://hdl.handle.net/2027.42/58243. Used with permission.)*

was consistently high in Aleppo province and in northwest Syria, at least in July. Looking at weather patterns over time, and comparing them to disease prevalence, should be informative, as should other factors associated with the risk and consequent spread of disease. Maps guide research; this sequence for July suggests forming similar sequences for each of the 11 other months.

Cutting across the Grain

The sequences in Figures 5.2 and 5.4 offered cuts of the temporal data 'orthogonal' to each other. These can be further compressed together using animation. Thus, a table containing the eight animations by year can offer the reader simultaneous views of the data by month (when hosted on a website with good connection speeds to ensure synchronization). Figure 5.5 shows a screenshot of the view of July, which is the same as the sequence of images in Figure 5.4. The time-spacing between successive frames of the annual animations is chosen to be the same from one animation to the next so that all are synchronized in appearance by month. The QR code in Figure 5.3 links to an animation that will show all months and flips through each month for each of the eight years, all within a single small frame.

In the case of the animation in Figure 5.5, the timing element is critical. Without author-created and -controlled synchronization, the animation simply becomes a hodgepodge of colored regions, imparting no sense of order or information to the observer.

Generally, provision for reader control (interactivity) with this sort of display might be an error and render the animation useless. There are, however,

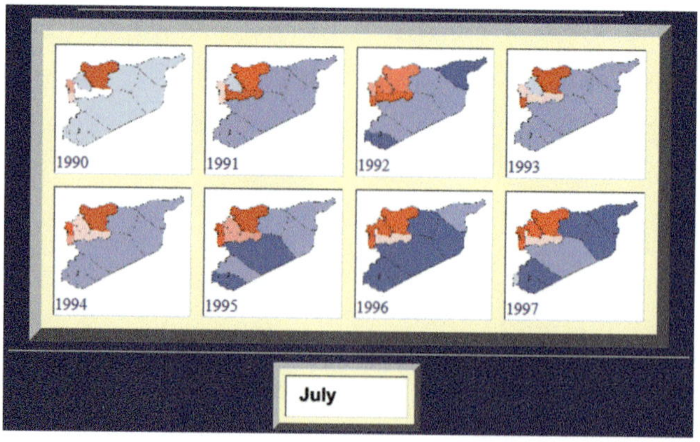

Figure 5.5. *Screenshot from an animated map sequence showing changing patterns of leishmaniasis incidence in July, over time, in Syria (by province). In the animated version, all 12 months appear. (Source: IMaGe, ©2002. http://hdl.handle.net/2027.42/58243. Used with permission.)*

special occasions on which limited user control might be appropriate. Consider the map sequence in Figure 5.4. If one wanted to spend time comparing and contrasting patterns between two consecutive Julys, flipping back and forth from one to the other, then providing user control of the animation would be important. Such provision is an easy matter: simply save the animation in a movie format in which the reader can mouse back and forth between successive animation frames.

Questions Generated by Mapping

Additionally, there are a number of content questions one might ask, based on observing this entire set of maps. If some of the questions have known answers, then this display might be calibrated as a 'model', after which one might then consider other questions with unknown answers. As examples, a few observations might be as follows:

- From 1995 on, the province of Damascus is always below the mean; prior to 1995, it was not and exhibited apparent seasonal variation, with values above the mean (for the most part) in October, November, December, January, and February. What did Damascus do in 1994–1995; was some sort of disease control measure implemented? If so, it may be working. What is the lesson, therefore, for Aleppo, which always appears above the mean? The controls applied in Damascus may require certain climatic/rainfall regimes or presence or absence of vegetation. Whatever the requirements, is the environment of Aleppo conducive to using the same sorts of control procedure that Damascus has employed?
- From 1995 on (and partly in 1994), provinces to the east of Aleppo begin to appear above the mean in a consistent pattern; why is this the case? The variation appears seasonal with high values in November, December, January, February, and March, and in that regard is similar to the pattern seen in Damascus (1990–1994); is that mere coincidence? What happened in 1994 to raise incidences to the east on an apparently persistent basis? Is there a relation to the Euphrates River Valley and to water projects to the north, in Turkey?
- Aleppo is almost always above the mean. The provinces to the west of Aleppo come in and out of the picture; is there some explanation for the pattern that appears?
- From 1995 on, Al Quneitera (the Golan Heights) appears not to be synchronized with the rest of the southern region as it had been before; why is this?
- The year 1994 seems a bit unusual, as if it were a transition point of some sort; what happened in 1994? Sometimes it appears to fit with the new grouping from 1995 on, and at other times it seems to fit with the old grouping from 1990–1993.

Haidar (2002) addresses complex questions such as these in her 2002 work. Animated maps that view spatial change over time can quickly generate new sets of questions derived from spatial thinking in environmental contexts.

References

Arlinghaus, S., S. Haidar, and M. Wilson. 2002. Animated map timeline, Syria. *Solstice: An Electronic Journal of Geography and Mathematics* 13(1).

Haidar, S. 1990–1997. Field data from Syria, by province.

Haidar, S. 2002. Environmental determinants of cutaneous leishmaniasis in Syria. PhD dissertation, University of Michigan, Ann Arbor. Persistent link: https://deepblue.lib.umich.edu/handle/2027.42/131932.

World Health Organization. 2013. Leishmaniasis. http://www.who.int/leishmaniasis/burden/magnitude/burden_magnitude/en.

6

Animap Abstraction: The Clickable Map, Virtual Reality, and More

Sandra L. Arlinghaus, with input from William C. Arlinghaus, Michael Batty, Klaus-Peter Beier, Matthew Naud, and John D. Nystuen

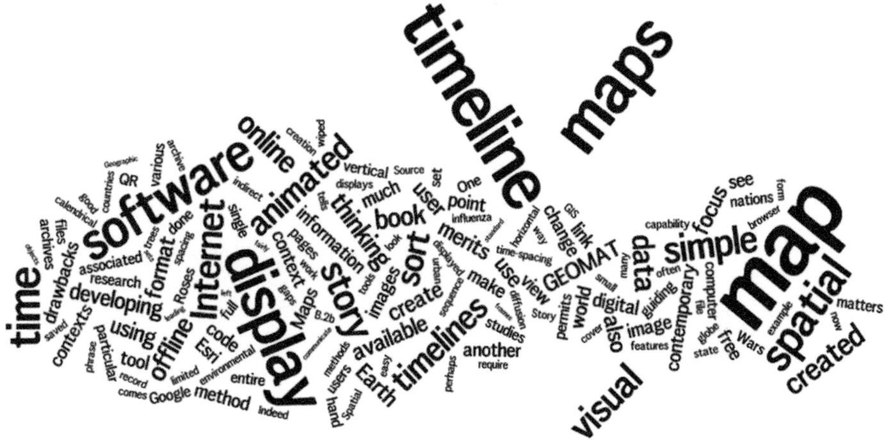

Figure 6.0 *Spatial word cloud summary, based on word frequency, made using Wordle.*

In the previous animation examples, broadly speaking, the final file was created by replacing, or overlaying, one map with another. To make the animation effective, the replacement/overlay process required careful attention to detail. Map boundaries from one map to the next needed to be precisely overlain, lest blurred images arise from sloppy boundary placement, introducing confusion in understanding content. So too, text from one map to the next needed to align exactly. With such precision in placement, only then would change from one frame to the next pop out at the reader. The reader would not know, from looking at the final animation, that it was composed of a sequence of full maps, identical for the most part across successive frames of the animation. Rather, it would appear to the reader that a single region, or a few regions, had been replaced from one frame to the next.

Hot Spot Map Regions: Varroa and Leishmaniasis

Software exists, and has for a long time, that permits the designer to note 'hot spots' as regions of change on a map (smaller than the entire map), and often, to make them clickable leading to images, webpages, or more. One popular tool for creating such hot spots is Dreamweaver; there are many others, some very specialized, as well. While hot spot creation is another way to work with maps, it generally lacks georeferencing except when it is part of geographic information system (GIS) software. The Varroa animation example might also be captured using an interactive hot spot map. Figure 6.1 shows such an example. The instances of Varroa are designated as hot spots, which can be clicked, on the map. A timeline feature lets the user choose the time period in which to look for interactive map hot spots. A new map at a more local scale pops up to show some of the hot spots. Maps of this sort can be quite useful, especially when coupled with animated .gifs, as in the Varroa maps in the earlier section. One gets a better idea of change over time with the simple animation and a better idea of detail with hot spots linked to extra materials. Indeed, one contemporary use of animation, tracking migratory bird populations, has readers calling for added hot spot capability within this fascinating application of animation with citizen science input (Leonard, 2016).

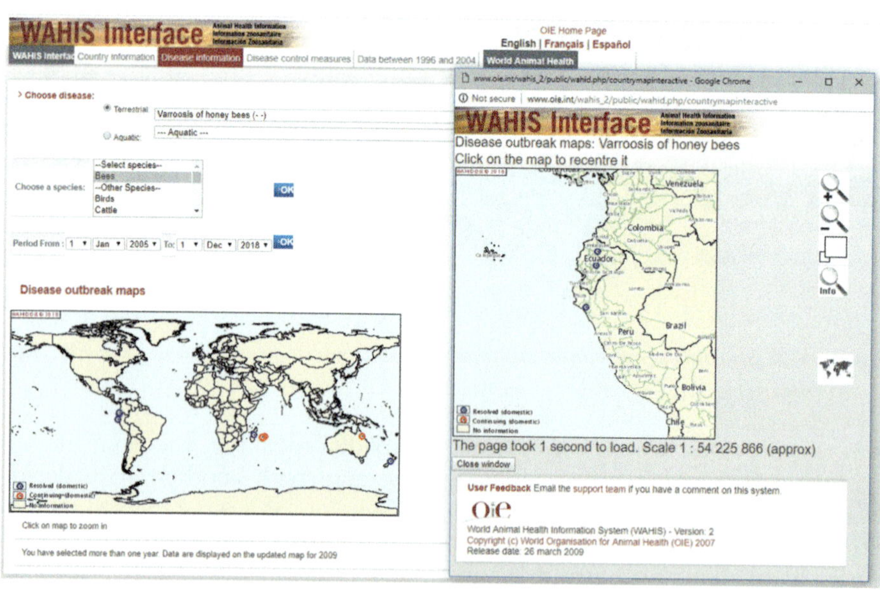

Figure 6.1 *Varroa hot spot–style map, with features involving an interactive timing element. (Source: World Organization for Animal Health, 2018. Reprinted with the kind authorization of the World Organization for Animal Health (OIE) (http://www.oie.int). Figure extracted from OIE website on December 7, 2018 at: http://www.oie.int/wahis_2/public/wahid.php/Diseaseinformation/Diseaseoutbreakmaps (a disease outbreak map for varroosis of honeybees as a base map overlain with a clicked-in enlargement of one region in western South America).)*

While the supplementation of an animated .gif with a hot spot map is helpful in the case of the Varroa study, it might not be as useful in the case of leishmaniasis in Syria (as presented here). In the latter case, the provinces of Syria had data associated with them that was analyzed, in advance of animation, within a GIS and then classified as being near, below, or above the mean. The fact that the underlying map collection was georeferenced to Earth coordinates was critical. Thus, the only hot spot maps that would offer good supplementation to those animated .gifs would be georeferenced ones created within a GIS, preferably within the same file that housed the existing analysis of the maps in the animated .gifs.

In the last step of creating an animation, the author has a choice in how to save the file, as a .gif (or similar) or as a .mov (or similar). The former keeps control in the hands of the author and not in the hands of the reader. The latter, however, offers interactive control by the reader. The decision rests on whether the gaps in the animation, set by the author, are important to the point the author is making and therefore worth controlling.

Clickable Maps

Animated maps function by replacing all, or part, of the map with another part of the map. If hot spots are used, links may be added to parts of a map. In addition, it is possible to add part of the underlying database, in a GIS, to a region in a map. Clicking on a region brings up the author-selected elements of the underlying database, as one form of a 'clickable' map. The clickable map in Figure 6.2 shows a screen capture of such a map, built on the City of

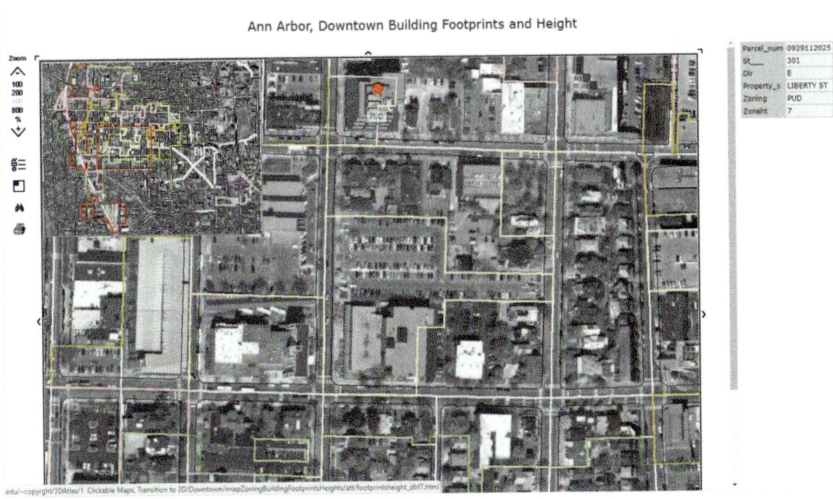

Figure 6.2 *Static image of a portion of a clickable map of Downtown Ann Arbor. (Source: IMaGe, ©2006. 3D Atlas of Ann Arbor, http://hdl.handle.net/2027.42/58271. Used with permission.)*

Ann Arbor parcel and zoning maps, together with an associated aerial of the region. Click on an outlined, zoned area and elements of the database pop up.

In Figure 6.2, an aerial file was inserted into Esri GIS software, which was then supplemented with ImageMapper software from Alta 4. This map offers readers the opportunity to click on a building in the aerial, surrounded by parcel boundaries from the official City of Ann Arbor parcel map, and see associated data set information; here, the author chose to have the fields showing parcel number, address, zoning type, and zoning height information displayed to the reader. The author retains control over what to show the reader. The reader has control over which parcels to choose and whether to display, in the upper-left corner, an inset map of the whole region while magnifying the map to get a better look. Here, a red dot (added in Adobe Photoshop) indicates which building was selected. When the parcel is moused over, a callout showing zoning for the parcel drops onto the map. When the associated parcel is clicked on, the corresponding data set entries pop up on the right. The beauty of this procedure is as follows:

- The digital divide is lessened, in certain ways.
  - There are plug-ins available for various versions of Esri software, old and new.
  - The clickable (interactive) map is created offline.
- Unlike many other hot spot maps, the map created here is georeferenced, so it can be used in conjunction with a host of other options that employ latitude and longitude, while it retains the advantages of a stick-in-the-pin map. It also deals swiftly with complex data sets and image sets because it is embedded in state-of-the-art GIS software.
- One issue with large clickable maps involves the transfer of data. While the file size is not large, there may be thousands, or tens of thousands, of very small files. Using ftp or similar software to transfer a very large number of files might be quite time-consuming.

Animation: Graph to Data Set Transformation and Vice Versa

Because GIS software provides seamless interactivity between maps and data sets, one might also wonder about making animated data sets: ones where a change in the data produces a change in the associated graph and vice versa. Figure 6.3 shows such an example using British population data based on rank differences over time. In it, the first data set shows data from 1880 highlighted among the set of all curves. As the animation changes, the dark line dances across the graph in synchronization with the dark box surrounding the column moving across the data set. The graph and the data sets are derived from complex relationships among rank–size issues and the associated population–environment dynamics of urban systems. Readers wishing to know more of those dynamics are invited to consult the original source

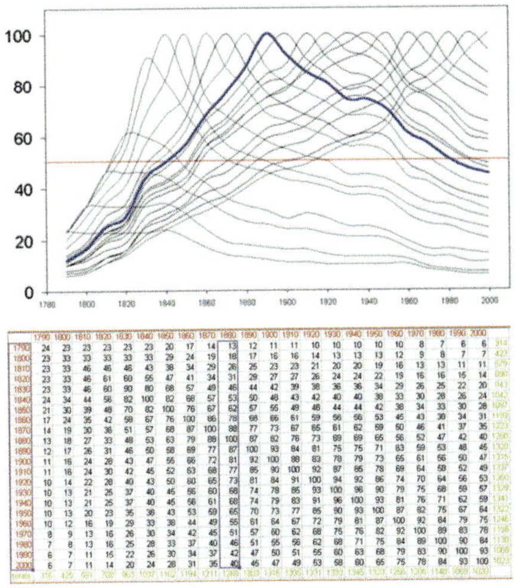

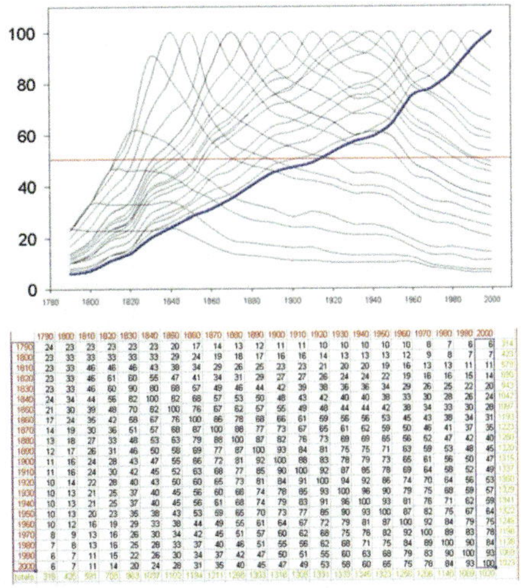

__Figure 6.3__ Synchronized graph and data set: animation. These screen captures show two frames from that animation: one associating the graph and data set from 1880 and the other showing that correspondence in 2000. The QR code provides a link to the animation. (Source: IMaGe, ©2003. http://hdl.handle.net/2027.42/58230. Used with permission.)

(Arlinghaus, Batty, and Nystuen, 2003); the point here is to demonstrate the possibility of a synchronized animation of graph and matrix columns.

The display in Figure 6.3 greatly facilitates disentangling pattern from a data set based on complex analysis; it also facilitates focusing on one curve at a time from a set of many. What it does not do, however, is what many animations fail to do: it is not georeferenced so that the reader does not know, at a glance, where the analysis is taking place.

Two- to Three-Dimensional Transformation

As 2D animation brought to life a 1D file of numbers, one might wonder if some form of 3D animation might inject life into 2D objects. In Figure 6.4, at the top, we see a static 2D image of a procedure for the fractal generation of a certain way of arranging hexagonal nets. While the overlays are interesting, suppose instead that that image is 'puffed up' as a 3D figure and colored with half-transparent colors and put into an interactive browser so that one can rotate it and look at how the different planes intersect or overlap each

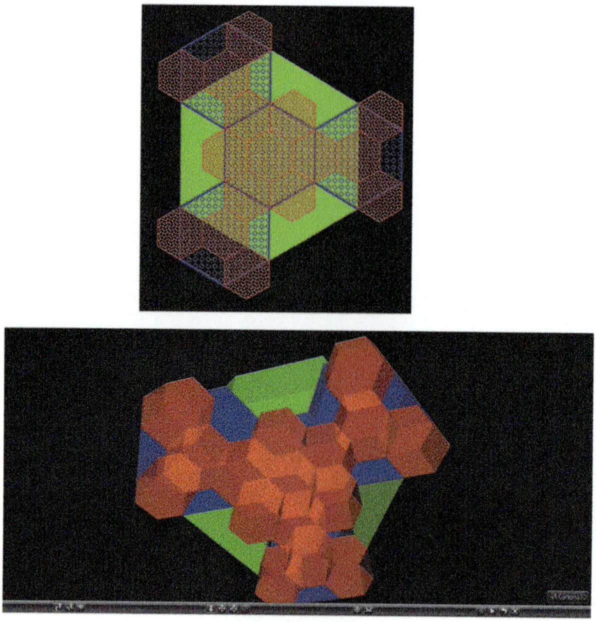

Figure 6.4 *Using the Cortona add-on to Internet Explorer to rotate and otherwise navigate the 3D image representation of the 2D conventional image. The QR code links to the file (http://www-personal.umich.edu/~copyrght/image/books/Spatial%20 Synthesis/tilek3transp3b.wrl). (Source: IMaGe, ©2005. Spatial Synthesis, Vol. I. http://hdl.handle.net/2027.42/58271. Used with permission.)*

other: to create a virtual reality within the static 2D figure. The procedure was executed using the Spatial Analyst extension to Esri ArcView to create vrml files to be loaded into an appropriate internet browser endowed with a plug-in (such as Cortona from ParallelGraphics) permitting the user to interactively navigate the virtual reality. It gives a new spatial view of that hexagonal environment and offers different directions for thinking, again helping to disentangle otherwise complex systems (Arlinghaus and Arlinghaus, 2005).

The Quest for Unification

Naturally, as one observes the power of various styles of animation and interaction, there is a desire to put it all together. The next case involves the possibility of flooding a creek near downtown Ann Arbor: the Allen Creek floodplain. Figure 6.5 shows a georeferenced location map for this creek, passing through the west side of Ann Arbor, in relation to standardly depicted urban features.

The Allen Creek floodplain was cast in a 3D model, built in 3D Studio Max software. Subsequent to that work, Naud reports that Paul Lippens created images for flood mitigation and put this original work into action (City of Ann Arbor, 2007, p. 18).

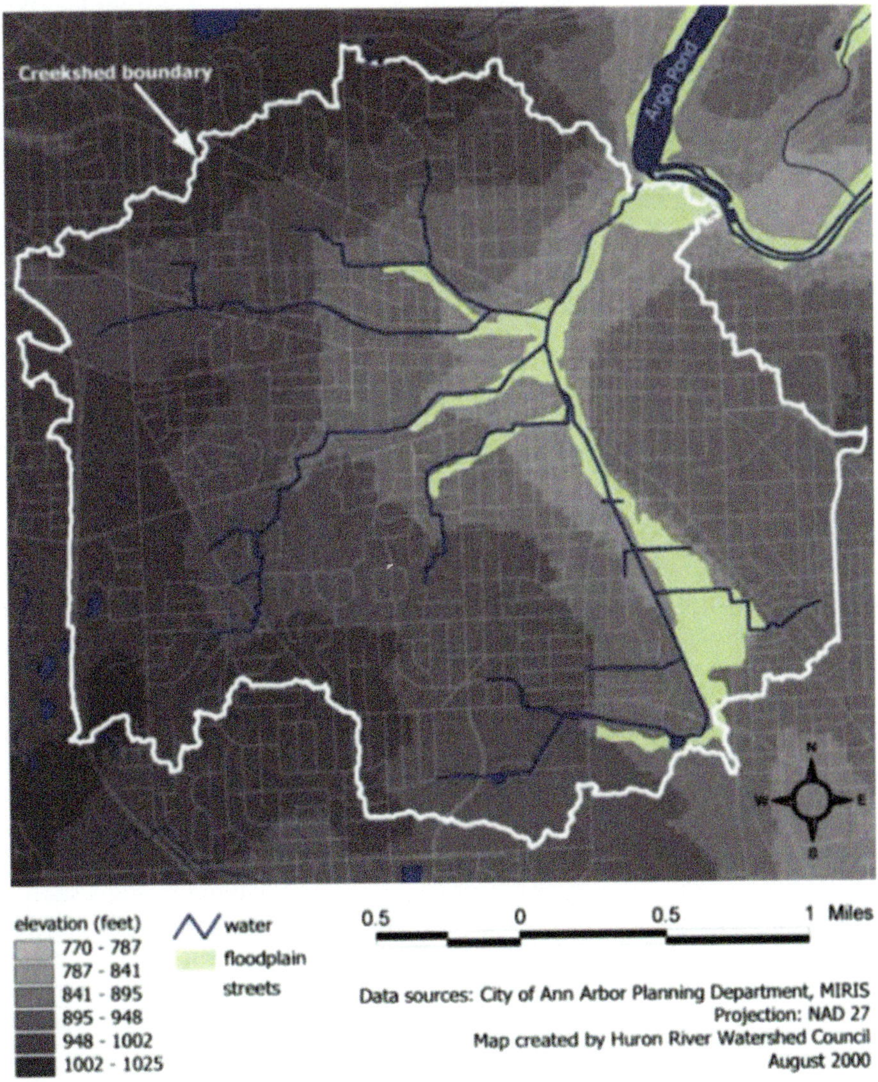

Figure 6.5 *Allen Creek and its floodplain cut a broad swath through the west side and center of Ann Arbor. (Source: City of Ann Arbor, public domain.)*

Beyond that, additional features were also incorporated into that virtual reality: items associated with human and commercial damage, possibly involving hazardous materials stored in buildings, a flood, street trees, buildings with and without textures, and street furniture. The inclusion of numerous worldly features made that model a very powerful display at the time of creation in the early 2000s . It was used in many different settings as a trigger to discuss planning in the downtown of Ann Arbor.

However, this model also failed to be georeferenced to the Earth. A 3D Cartesian coordinate system is embedded into 3D Studio Max, so any terrain measurements made using contours are made with respect to contours in the internal environment of 3D Studio Max rather than to contours with Earth coordinates.

Figure 6.6 shows a sequence of three images created in 3D Studio Max, incorporating the floodplain, 3D buildings, and various other features. A set of QR codes links the reader to the associated virtual reality files to open in a browser with Cortona (or similar). These virtual reality files are far more complex than one might imagine from simply looking at the images.

Each static image in Figure 6.6 represents a virtual reality file in which one can drive, fly, or walk around and look at the various features within the 3D environment. The three files are distinct but linked to each other; each contains background music suited to the depicted environment. In the first file, the top image in Figure 6.6, one sees picturesque downtown Ann Arbor, looking west from Main Street and Liberty Street into the Allen Creek floodplain. It's a beautiful day for a walk while listening to the strains of Beethoven's Pastorale Symphony interspersed with the happy sounds of voices enjoying sidewalk eateries and other pleasant downtown amenities. The second file, represented by the screen capture in the middle frame of Figure 6.6, begins with the gathering storm movement of Beethoven's Sixth Symphony under gray skies. An explosion in the background represents the strain of the rain storm on nearby Argo Dam with the subsequent rise in creek waters in Allen's Creek. The viewer sees, through a sequence of eight subordinate virtual reality files embedded within a parent file, the rising animated creek as it engulfs nearby buildings and parcels (using the internal coordinate system of 3D Studio Max). We hear the sounds of emergency vehicles and the shrieks and cries from the nearby residential neighborhoods. The red 'chimneys' on the roofs of commercial buildings are author-added 'handles' for links to html files that inventory the content of the buildings under them (a suggestion of Matthew Naud, who at the time was also the emergency manager for the City of Ann Arbor). Click on a chimney and go to the associated file (one such html file is shown in Figure 6.7) for that building. In the file associated with the third image in Figure 6.6, the floodplain is littered with wreckage from the disaster; the Lacrimosa from Mozart's Requiem Mass plays in the background. This set of files, while not an accurate depiction of flooding possibility, was a useful tool for prompting public interest in matters associated with the Allen Creek floodplain and possible development projects.

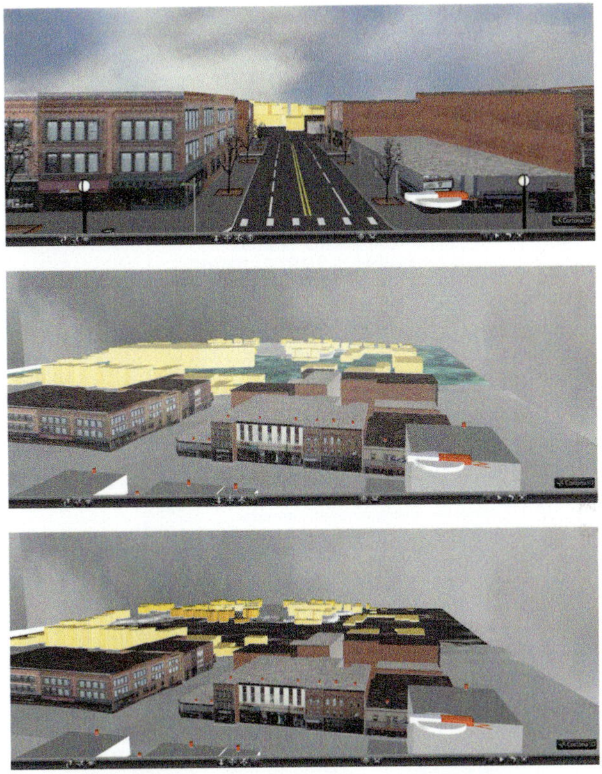

(a) (b) (c)

Figure 6.6 *Virtual reality of downtown Ann Arbor in association with flooding of Allen Creek and with a data set inventorying hazardous material locations within buildings. Use Cortona, associated with Internet Explorer (works in Windows 10) to view the virtual reality file associated with each discrete image above. (Source: Arlinghaus, 2005. Used with permission. QR codes that link to 3D models: (a) http://www-personal.umich.edu/~copyrght/Archimedes/FLOOD01.wrl; (b) http://www-personal.umich.edu/~copyrght/Archimedes/FLOOD02.wrl; (c) http://www-personal.umich.edu/~copyrght/Archimedes/FLOOD03.wrl. Images reprinted with permission of the Institute of Mathematical Geography, http://www.imagenet.org).)*

Also lurking behind the flooding scene of downtown by Allen Creek are substantial data sets of code. Open the .wrl (vrml 97) file in Notepad++ or similar to see the more than 150,000 lines of code that creates each of the three files associated with the screen captures in Figure 6.6. Figure 6.8 shows a few lines of code associated with one of the files. Notice the

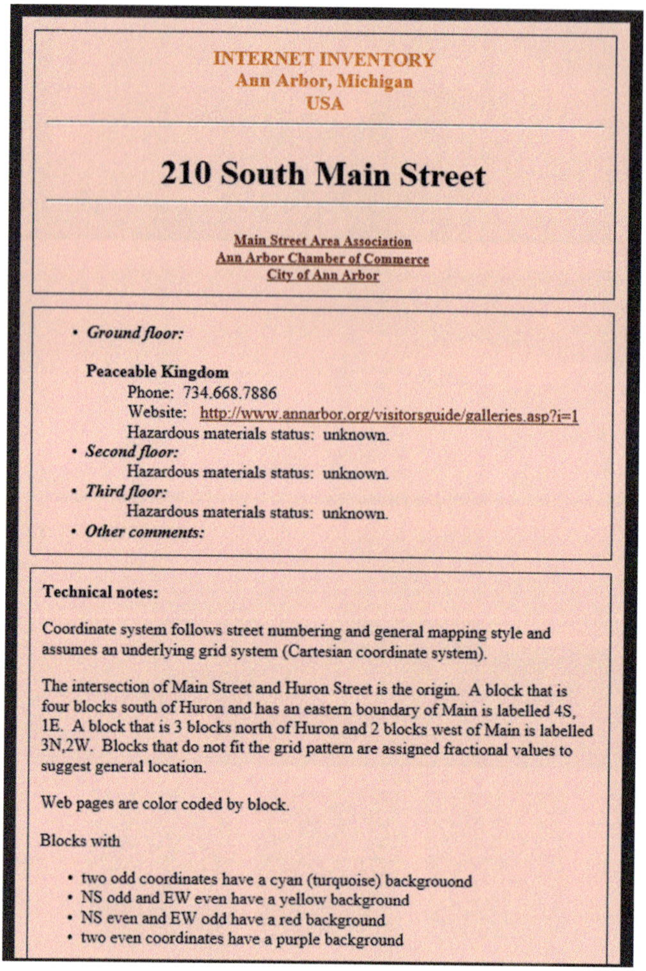

INTERNET INVENTORY
Ann Arbor, Michigan
USA

210 South Main Street

Main Street Area Association
Ann Arbor Chamber of Commerce
City of Ann Arbor

- *Ground floor:*

  Peaceable Kingdom
  Phone: 734.668.7886
  Website: http://www.annarbor.org/visitorsguide/galleries.asp?i=1
  Hazardous materials status: unknown.
- *Second floor:*
  Hazardous materials status: unknown.
- *Third floor:*
  Hazardous materials status: unknown.
- *Other comments:*

Technical notes:

Coordinate system follows street numbering and general mapping style and assumes an underlying grid system (Cartesian coordinate system).

The intersection of Main Street and Huron Street is the origin. A block that is four blocks south of Huron and has an eastern boundary of Main is labelled 4S, 1E. A block that is 3 blocks north of Huron and 2 blocks west of Main is labelled 3N,2W. Blocks that do not fit the grid pattern are assigned fractional values to suggest general location.

Web pages are color coded by block.

Blocks with

- two odd coordinates have a cyan (turquoise) backgrouond
- NS odd and EW even have a yellow background
- NS even and EW odd have a red background
- two even coordinates have a purple background

Figure 6.7 *Click on the red 'chimney' on 210 S. Main Street to bring up an inventory geared toward hazmat awareness. (Source: IMaGe, ©2005. http://www.imagenet.org. Used with permission.)*

reference to textures linked via a URL. These textures are images applied to underlying plain, chunky buildings, such as those yellow buildings seen in the distance in Figure 6.6, to make them look like actual buildings. They were acquired by going to the downtown and using a camera on a tripod to photograph building façades at a consistent distance from the building. The photographs were then brought back to a computer laboratory at the University of Michigan, where the trapezoidal shapes, which necessarily result because the vantage point of the camera converts the rectangular shape of a tall building to a trapezoid, were converted back to rectangles. After that, painstaking removal of the foreground (shrubs, passing cars, pedestrians) ensued and was then followed by using the 'clone stamp' tool to replace the foreground with imagery that made the building look 'whole' once again (using images of pieces of window framing, bricks, and so forth). Patterns that seemed reasonable were applied to unphotographed parts of buildings. The tools used were primarily from Adobe Photoshop. Once the textures were applied, and the model fairly complete, the model was taken

```
  ]
}
DEF Base32 Transform {
  translation 513 394 -95.4
  rotation 0 -0.707 0.707 -3.14
  children [
    Shape {
      appearance Appearance {
        material Material {
          diffuseColor 0.588 0.588 0.588
          ambientIntensity 1.0
          specularColor 0 0 0
          shininess 0.145
          transparency 0
        }
        texture ImageTexture {
          url ["C_10_N.jpg"]
        }
      }
      geometry DEF Base32-FACES IndexedFaceSet {
        ccw TRUE
        solid TRUE
        convex TRUE
        coord DEF Base32-COORD Coordinate { point [
          34 0.909 7.06, 65.1 0.52 7.06, 34 0 -6.25, 65.1 -0.389 -6.25]
        }
        coordIndex [
          0, 1, 3, 2, -1]
        texCoord DEF Base32-TEXCOORD TextureCoordinate { point [
          0.000499 0.0005, 1 0.000499, 0.0005 1, 1 0.999]
        }
        texCoordIndex [
          0, 1, 3, 2, -1]
        }
    }
  ]
}
DEF Base33 Transform {
  translation 752 402 -112
  rotation 0 -0.707 0.707 -3.14
  children [
    Shape {
      appearance Appearance {
        material Material {
          diffuseColor 0.588 0.588 0.588
          ambientIntensity 1.0
          specularColor 0 0 0
          shininess 0.145
          transparency 0
        }
        texture ImageTexture {
          url ["C_01_N.jpg"]
        }
      }
      geometry DEF Base33-FACES IndexedFaceSet {
        ccw TRUE
        solid TRUE
```

Figure 6.8 *A sample of the more than 150,000 lines of code creating the image file in Figure 6.6.*

downtown and compared with the actual 'real' scene (as strongly encouraged by Peter Beier). Adjustments were made and street furniture, created by students on the team the author directed as faculty advisor in Engineering 477, was inserted in the first file (Domzal, Hwang, and Walters, 2005). Street textures created by Peter Beier (2004) were inserted as well.

Following the creation of these files and the field testing of them from the mechanical standpoint, it was time to take them out to the municipal environment, where they were displayed and commented on by a variety of groups interested in downtown issues and in floodplain issues, especially in association with Allen Creek (Figure 6.5). Special presentations were given to members of the City of Ann Arbor Planning Commission, City Council, Environment Coordination Services, University of Michigan groups at the 3D Laboratory, and the Allen Creek Watershed Group. See the preface to the book for details. Judy McGovern from *Ann Arbor News* summarized the events in an article at the time (Sunday, November 27, 2005).

> I happened to call Sandy Arlinghaus when, as luck would have it, our minds were on the same thing. Well, more or less the same thing. What a coincidence, the former chairwoman of the city Planning Commission said, "I'm flooding Ann Arbor right now." Now, this particular flood was confined to a computer model running in Arlinghaus's office in the University of Michigan's School of Natural Resources and Environment. But Arlinghaus, who teaches mathematical geography and population–environment dynamics at U-M, worries that in the future streets and neighborhoods could be really, rather than virtually, swamped.

As McGovern put it in the context of the possible development of the time involved in remaking the downtown, "Stay tuned." Stay tuned here, though, for a variety of methods to introduce even more visual features and interactivity, such as 3D reliefs and the georeferencing of materials. All in the quest to use spatial thinking to generate optimal organization for any project at hand, using maps, archives, and timelines.

References

Arlinghaus, S. L. 2005. Archimedes in Ann Arbor. *Solstice: An Electronic Journal of Geography and Mathematics* 16(2). Retrieved from http://www-personal.umich.edu/~copyrght/image/solstice/win05/3DatlasF05.html.

Arlinghaus, S. L., and W. C. Arlinghaus. 2005. *Spatial Synthesis, Volume 1, Book 1: Centrality and Hierarchy*. Ann Arbor, MI: Institute of Mathematical Geography, University of Michigan. Retrieved from http://www-personal.umich.edu/~copyrght/image/books/Spatial%20Synthesis2/1ndex.htm.

Arlinghaus, S. L., M. Batty, and J. D. Nystuen, with input from N. Shiode. 2003. Animated time lines: Coordination of spatial and temporal information. *Solstice: An Electronic Journal of Geography and Mathematics*. Retrieved from http://www-personal.umich.edu/~copyrght/image/solstice/sum03/batty.html.

Arlinghaus, S. L., et al. 2006. *3D atlas of Ann Arbor,* 1st edition. Retrieved from http://www-personal.umich.edu/~copyrght/3DAtlas/3dAtlasFrameset.htm.

Beier, K.-P. 2004. One optimization of an earlier model of virtual Downtown Ann Arbor. *Solstice: An Electronic Journal of Geography and Mathematics* 15(2). Retrieved from http://www-personal.umich.edu/~copyrght/image/solstice/sum04/beieredited/beier.html.

City of Ann Arbor. 2007. Flood mitigation plan. https://www.a2gov.org/departments/systems-planning/planning-areas/water-resources/floodplains/Documents/FloodplanMitigationPlan_Mar07.pdf

Domzal, A. J., U. S. Hwang, and K. J. Walters, Jr. 2005. Virtual flood in the Allen Creek floodplain and floodway. *Solstice: An Electronic Journal of Geography and Mathematics.* 15(2). Retrieved from http://www-personal.umich.edu/~copyrght/image/solstice/win05/3DatlasF05.html.

Leonard, P. 2016. Mesmerizing migration: Watch 118 bird species migrate across a map of the Western hemisphere. Cornell Lab of Ornithology. Retrieved from https://www.allaboutbirds.org/mesmerizing-migration-watch-118-bird-species-migrate-across-a-map-of-the-western-hemisphere.

McGovern, J. 2005. Whose vision will prevail in Ann Arbor? *Ann Arbor News,* November 27.

World Organization for Animal Health. 2018. WAHIS Interface, Version 1. Retrieved from http://www.oie.int/wahis_2/public/wahid.php/Diseaseinformation/Diseaseoutbreakmaps?disease_type_hidden=&disease_id_hidden=&selected_disease_name_hidden=&disease_type=0&disease_id_terrestrial=125&disease_id_aquatic=-999&speciesselect%5B%5D=1&selected_start_day=1&selected_start_month=1&selected_start_year=2005&selected_end_day=1&selected_end_month=12&selected_end_year=2018&submit2=OK.

III

3D Maps: Georeferencing, Turn of the Millennium

This section begins with a view of the Varroa study in Google Earth. That method involves a contemporary view of the time to contrast with simple animated mapping. With Google Earth, one can see terrain and other features associated with the spread of the mite; such views were not available in the animap. However, the animap covered the full globe in a single view, which is not possible in Google Earth. The section continues to look at other environmental contexts in Google Earth, often combining animated maps with the Google Earth views.

7

3D Maps: Varroa Honey Bee Mite and More

Sandra L. Arlinghaus and Diana Sammataro

Figure 7.0 *Spatial word cloud summary, based on word frequency, made using Wordle.*

3D Varroa

The animated Varroa map (Figure 7.1) is built on a simple line base map. Thus, one cannot look for patterns of Varroa distribution in conjunction with underlying physical or human geographic patterns on the surface of the Earth. When dealing with biological organisms, such associations may be important. Once Google Earth became available, Arlinghaus suggested and implemented the placing of information from the animated map into the virtual world of the Google Earth globe, which has the capability to show terrain and a host of other geographical features together with the Varroa mite data from the animated map (Arlinghaus and Sammataro, 2009). Figure 7.1 shows a screen capture of some of that file.

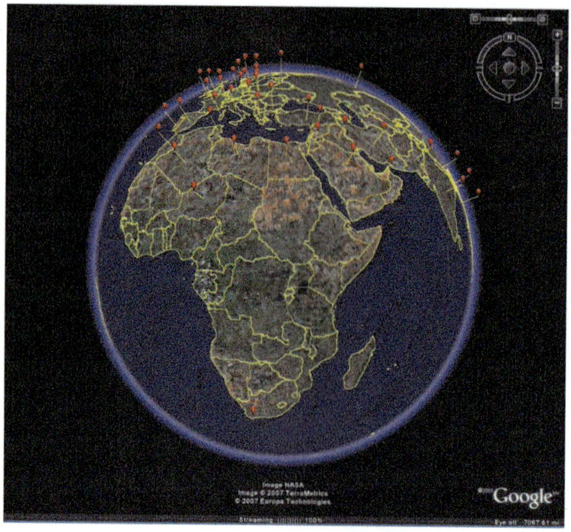

Figure 7.1 *Red push-pins on the globe show Varroa mite presence. The table at the left interacts with the globe and shows year and country names. (Source: IMaGe, ©2009. Used with permission.)*

Figure 7.2 *QR code link to .kmz file of Varroa data sets to open in Google Earth. (Source: http://www.mylovedone.com/image/solstice/win11/varroa.jpg.)*

To see a full image and to manipulate the globe itself in association with the full database, the QR code in Figure 7.2 will send the reader to the .kmz file to open in Google Earth.

Google Earth views are limited because less than half the Earth is visible at a time. The capability to show terrain and various other elements of the physical and human geographic landscape is, however, a large addition, as are the capabilities to zoom in and out, rotate the Earth sphere, and manipulate default software layers and add new ones. Google Earth views of the Varroa mite diffusion process are fine supplements to illustrate processes that can be contained within a less than global viewpoint and that benefit from the simultaneous display of underlying terrain and physical geography.

Combining the Local with the Global

Another advantage that the Google Earth display has is that it is interactive. Thus, one can consider mapping local data from field evidence at different time periods and then uploading it into an existing file that is more global. The interactivity creates the opportunity for seamlessly displaying different scales of data at different times. Indeed, as time passes, related (but different) interests might be included as well. We illustrate all of these ideas with the inclusion of an example from earlier work (pre-2015) that incorporates these capabilities.

The Case of Hawaii

Another pest that may appear in geographical association with honeybees is the small hive beetle. In 2010 and later, data was becoming available to Sammataro about the prevalence of this pest in Hawaii. Figure 7.3 shows one frame of an animation created in Google Earth that indicates the extent of prevalence of the pest. In the animation, a sequence of images of the temporal documenting of the pest culminates in the image in Figure 7.3.

Global datasets on the spread of a pest are typically compiled over time from regional sightings at a local geographical scale. Such datasets often become available through direct communication in advance of their formal publication. Field study, travel, and conferences, with appropriate email follow-up, become important ways to acquire timely information. The interactivity of Google Earth can serve as a simple yet powerful means

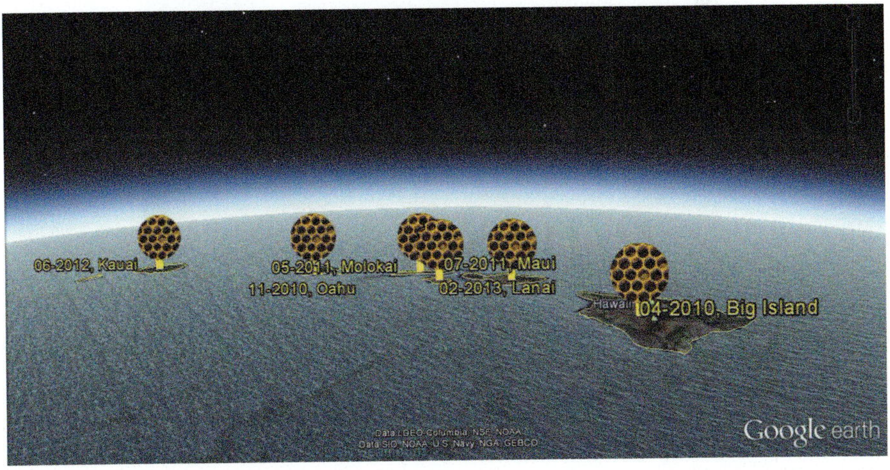

Figure 7.3 *Single frame from an animation showing the spatial extent of data recording the small hive beetle on the Hawaiian Islands before 2015. (Source: IMaGe, ©2016. Used with permission.)*

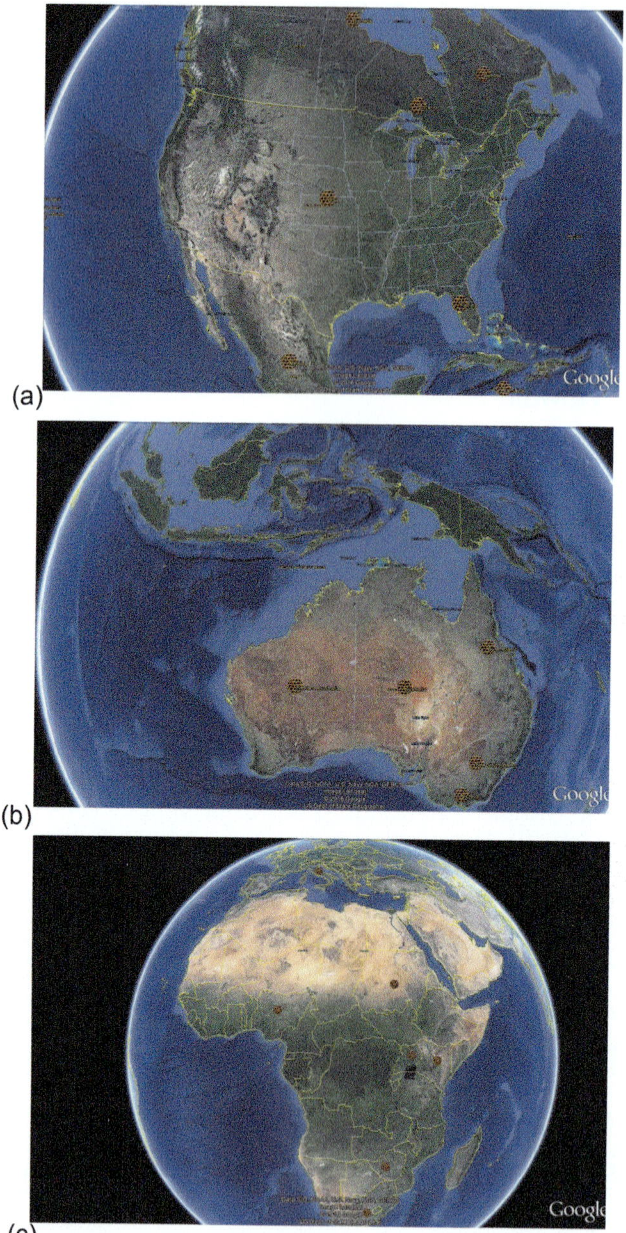

Figure 7.4 *Three views of a globe file showing sites where small hive beetle data were collected. Spin it around to Hawaii to also see the view from Figure 7.3. All collected in a single image. (Source: animation and .kmz/.kml file, IMaGe, ©2016. Used with permission.)*

for augmenting and even leading such communication as it unfolds. Once mapped, even in the field on a laptop, the dataset can later be updated or changed back in the lab.

Online Sources

Online datasets are often prevalent and helpful in understanding spatial patterns. However, their different formats can be daunting to try visualizing varied sets. The interactivity of Google Earth to present datasets at various geographical scales is particularly useful. In the case of the small hive beetle, it is natural to ask where else in the world, in addition to Hawaii, one might find evidence of it. The question of how pests get to remote islands always makes for interesting speculation.

We coupled data sources from the University of Florida (Ellis and Ellis, 2010) with others from the Invasive Species Compendium (2012) and the Ontario Ministry of Agriculture, Food and Rural Affairs (2011). Then, we visualized them in Google Earth to create an animation that showed, in a single moving image, where the small hive beetle might be found across the globe together with the more local data set from Hawaii. Unity across different geographical scales was created using a custom place mark showing a circular swatch of honeycomb (Sammataro and Arlinghaus, 2016). Figure 7.4a–c shows a few frames from that animation, which attempts to integrate the global with the local. Simply spin the globe and see the broad or local picture.

Future Direction

The mission of this chapter was twofold. First, it was to compare and contrast two different mapping tools using the same environmental context: using animated mapping and interactive 3D mapping on issues associated with the global honeybee population. Second, it was to illustrate the importance of interactivity in being able to combine data sets of varying scales and formats and to do so in a way that makes the updating of research a dynamic process. The rest of the section will show the uses of 3D mapping, perhaps coupled (as was the case here) with animation.

References

Arlinghaus, S. L., and D. Sammataro. 2009. Bee Ranges and almond orchard locations: Contemporary visualization. *Solstice: An Electronic Journal of Geography and Mathematics* 20(1).

Ellis, J. D., and A. Ellis. 2010. Featured creatures, University of Florida, Entomology & Nematology.

Invasive Species Compendium. 2012.
Ontario Ministry of Agriculture, Food and Rural Affairs. 2011. Ontario provincial apiarist annual report.
Sammataro, D., and S. L. Arlinghaus. 2016. Small hive beetle animaps: Focus on Hawaii. *Solstice: An Electronic Journal of Geography and Mathematics* 27(1).

8

3D Charts: Greater London, 1901–2001

Georeferencing of Population Data and Rank–Size Patterns

Michael Batty and Sandra L. Arlinghaus

Figure 8.0 *Spatial word cloud summary, based on word frequency, made using Wordle.*

Greater London: A Century of Change

Greater London is composed of the City of London and 32 boroughs that surround the central city. We begin by looking at changes in rank and size in the data sets of interest, by decade, over the course of the 20th century. The concepts of rank and of size, and their relative relationships, are important because position can affect the funding of government and other projects, population patterns and clustering that can bring about economies of scale for investment in projects, and various other population–environment considerations. Imaginative views of rank and size, and their relationships, is a subject of continuing interest (Batty, 2006). We look initially at patterns for Greater London and then focus our scale of observation more locally, on the boroughs composing Greater London.

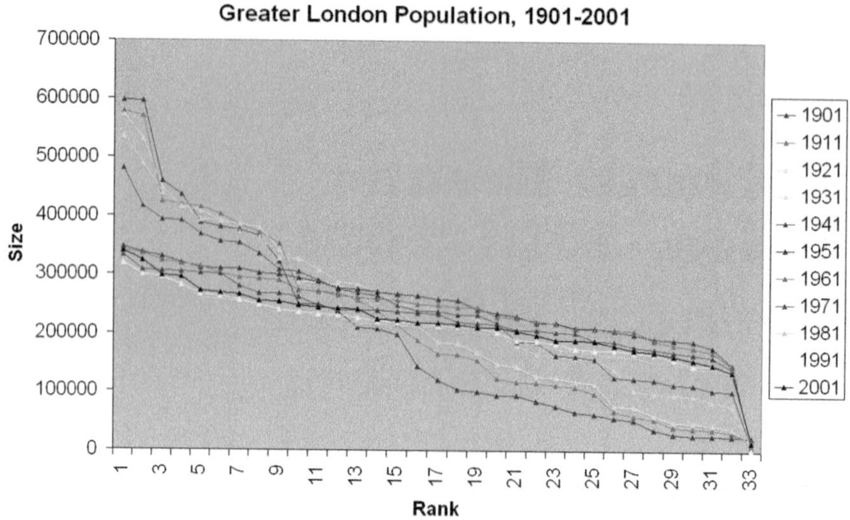

Figure 8.1 *Rank–size plots of the City of London and the 32 surrounding boroughs composing Greater London. (Source: IMaGe, ©2006. http://hdl.handle.net/2027.42/58245. Used with permission.)*

Rank–Size Plots

Rank–size plots for Greater London and 32 surrounding boroughs are shown in Figure 8.1. In the timeline on the right, one sees that the steepest pattern from high- to low-ranking cities appears in 1901. In the decades from 1901 to 1941, the steep pattern remains, although the degree of steepness diminishes as time progresses. From 1951 onward, the pattern markedly flattens out; the size of the 25th ranked city/town (for example) is no longer as distant from the top-ranked cities as it was earlier. There appears to be a change in pattern around the time of World War II, presumably with people moving their residences out of a highly concentrated central city to surrounding territory. The graph, however, gives no visual evidence to support this inference from the graph.

Mapping a Graph: Georeferencing of Data for the Region

Graphs such as the one in Figure 8.1, and the animated one from Chapter 6, are those traditionally associated with displaying the ranks and sizes of cities. Animating them is more recent and permits the compression of a large amount of information in one image. Neither approach, however, displays the information in a georeferenced manner. If we think of the graph in Figure 8.1 as a bar chart, and then convert those bars to 3D, it is possible to georeference the data by placing the 3D bar of data of a given borough on its

Figure 8.2 *Greater London, 1901. Red 3D bars represent total population per borough. (Source: IMaGe, ©2006. http://hdl.handle.net/2027.42/58245. Used with permission.)*

appropriate position on a map. To execute the process, we employed Google Earth for georeferencing on a sphere, and Google (now Trimble) SketchUp to create the 3D bar, suitably georeferenced, and then exported the georeferenced object back to Google Earth. Figure 8.2 displays the pattern of georeferenced data for 1901; Figure 8.3 displays the corresponding pattern for 1951; and, Figure 8.4 displays the associated pattern for 2001. We have similar images for each decade from 1901 to 2001 animated in a single file. In these images, the taller the 3D bar, the higher the total population for a particular borough.

Notice that in 1901, the population is heavily concentrated in the central core. By 1951 there is substantial decline in the core, with the distribution more evenly spread across the entire region. In 2001, the central core appears depressed in height in relation to the surrounding suburban parallelepipeds. Our guesses from looking at the data in Figure 8.1 are confirmed by the evidence of georeferencing the data sets in maps. The QR codes in Figure 8.5a,b give direct links to selected animated files associated with the text; additional material appears in an article cited in the references (Arlinghaus and Batty, 2006).

If one asks when this shift might have happened, and studies the entire sequence, it appears that post–World War II marks the beginning of this shift (see the 1951 frame). Again, the map sequence offers evidence to guide a

Figure 8.3 *Greater London, 1951. Red 3D bars represent total population per borough. (Source: IMaGe, ©2006. http://hdl.handle.net/2027.42/58245. Used with permission.)*

Figure 8.4 *Greater London, 2001. Red 3D bars represent total population per borough. (Source: IMaGe, ©2006. http://hdl.handle.net/2027.42/58245. Used with permission.)*

Figure 8.5 *(a) Greater London .kmz file, see download files in references. (b) Animated version of Figure 8.1, see download files in references.*

research study to consider the shift in population in Greater London caused by World War II and various associated issues. The entire set of georeferenced data by decade may also suggest, especially to those quite familiar with British history, other patterns subtler than this one.

It might be interesting to compare and contrast this situation in London with other major cities, both in the United Kingdom and elsewhere, especially in regard to movement patterns, and sprawl, in relation to war. Indeed, one might consider applications of this method to other urban areas in order to study land use planning, circulation, and infrastructure in relation to disasters.

Local Focus

The georeferenced images in Figures 8.2, 8.3, and 8.4 offer views of data not present in Figure 8.1. A limitation of Figures 8.2, 8.3, and 8.4 involves the clustering of the vertical bars. While the clustering presents evidence of a general pattern for the entire region over time, it is difficult to disaggregate the bar for a single borough and look at it over time. Adjustment in scale of study, from regional to local, might be appropriate when studies of population change over time are related to fiscal issues for local municipalities.

Because the maps for the region were composed of individual 3D bars, it is a simple matter to remove bars from the animation and make comparisons, as desired. For example, the borough of Tower Hamlets is adjacent to the City of London: it is a 'close-in' borough. Figures 8.6a, 8.7a, and 8.8a make it easy to compare and contrast the relative rise and fall in population of Tower Hamlets over the century from 1901 to 2001 with 1951 as an appropriate postwar midpoint. Figures 8.6b, 8.7b, and 8.8b provide a similar display for Barnet, a 'far-out' borough, over the same time span. And, in terms of visualizing possible urban sprawl, it is easy to look at simultaneous patterns of rises and falls in population in relation to distance from the core, by looking at the sets of three pairs: Figure 8.6a,b, 8.7a,b, and 8.8a,b, also over the same time span.

Generally, scholars investigating patterns associated with sprawl might find this sort of tool helpful in a variety of settings. In the context of this work, what is important is the use of bar charts that are georeferenced so that one might see complex quantitative population–environment data linked with geographic position, terrain, and so forth, as well as with time span.

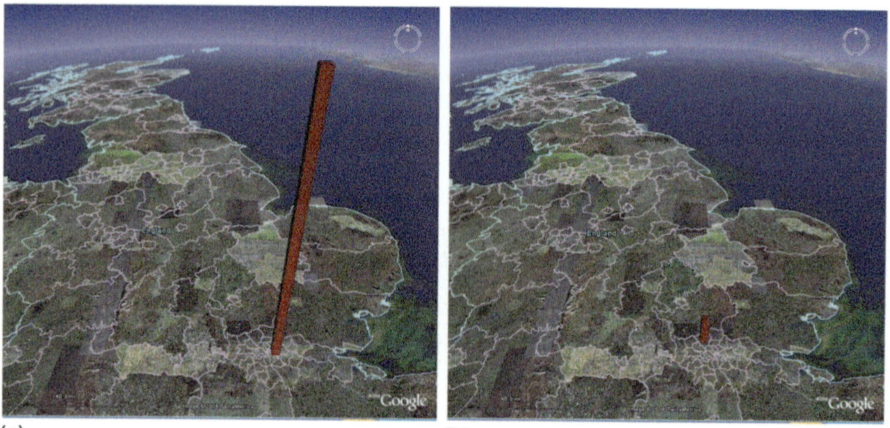

(a) (b)

Figure 8.6 *1901. (a) Tower Hamlets left. (b) Barnet right. (Source: IMaGe, ©2006. http://hdl.handle.net/2027.42/58245. Used with permission.)*

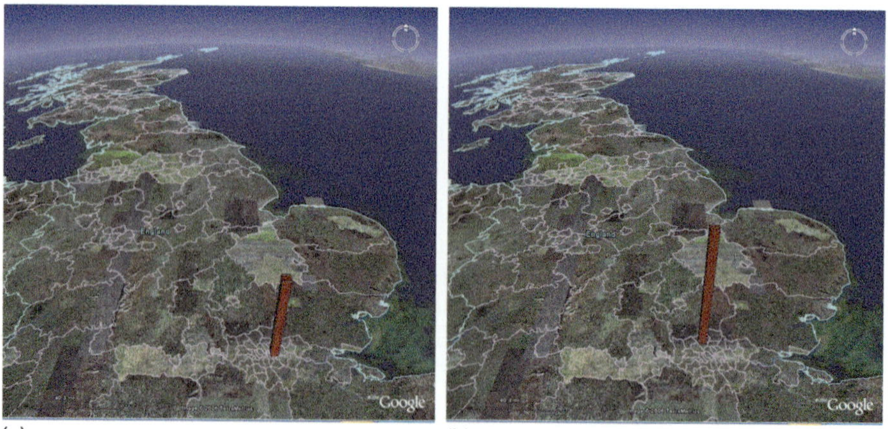

(a) (b)

Figure 8.7 *1951. (a) Tower Hamlets left. (b) Barnet right. (Source: IMaGe, ©2006. http://hdl.handle.net/2027.42/58245. Used with permission.)*

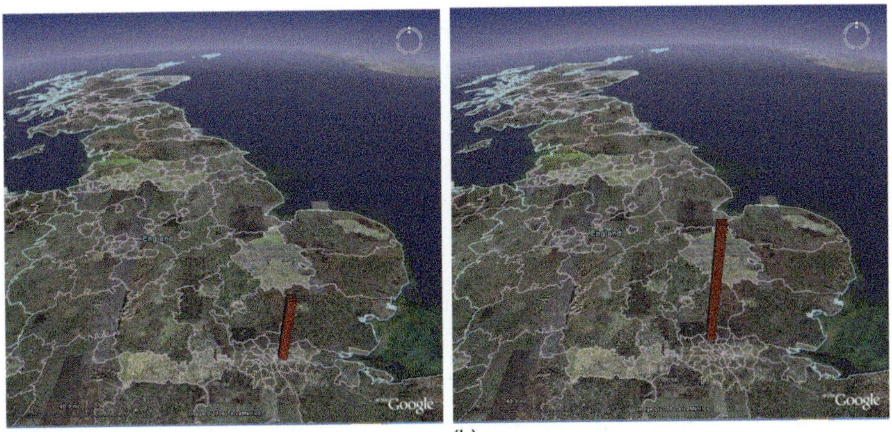

(a) (b)

Figure 8.8 *2001. (a) Tower Hamlets left. (b) Barnet right. Tower Hamlets close to core; Barnet away from Core. (Source: IMaGe, ©2006. http://hdl.handle.net/2027.42/58245. Used with permission.)*

Download Files

Figure 8.5a: Greater London, full .kmz file of population–environment visualization (http://www-personal.umich.edu/%7Ecopyrght/image/solstice/win06/arlbat2/GreaterLondon.kmz).

Figure 8.5b: Animated version of Figure 8.1 (http://www-personal.umich.edu/%7Ecopyrght/image/solstice/win06/arlbat2/GreaterLondon.gif).

References

Arlinghaus, S., and M. Batty. 2006. Visualizing rank and size of cities and towns, part II: Greater London. *Solstice: An Electronic Journal of Geography and Mathematics* 17(2). Special issue on geometry and the internet. Retrieved from http://www.imagenet.org; http://www-personal.umich.edu/~copyrght/image/solstice/win06/arlbat2/indexPartII.html.

Batty, M. 2006. Rank clocks. *Nature* 444, 30 November, 2006. doi:10.1038/nature05302.

9

3D Charts: Air Pollution Changes in Metropolitan Detroit, 1988–2004

Kerry Ard

Figure 9.0 *Spatial word cloud summary, based on word frequency, made using Wordle.*

Introduction

Google Earth can be an effective tool to educate the public about environmental risk, especially when it is coupled with the Environmental Protection Agency (EPA)'s Risk Screening Environmental Indicators (RSEI) data. This data set employs information submitted by facilities that emit toxic pollutants into the air, such as chemicals distributors, manufacturers, mining industries, utility operations, and hazardous waste treatment and disposal plants. The toxicity of over 600 individual chemicals, amounts released, and the path the chemical is expected to take as it is disbursed into the envirnment (determined by weather patterns) is modeled and a unitless score is created so that one might make comparisons of health risks. This risk score allows one to see which geographic areas have higher amounts of airborne toxic chemicals. This project animates the changes in the amounts of these chemicals for each

1×1 km^2 grid cell within three Michigan counties: Wayne, Oakland, and Macomb, for the years 1988–2004.

The publicly available versions of these data are provided on a facility-by-facility basis. However, this project utilizes data provided by Abt consultants, a group hired by the EPA to use air dispersion modeling to estimate where facility-emitted air pollutants can be expected to be found around EPA-regulated facilities annually.

Model

The base from which to generate the model in Figure 9.1 was developed by the author in association with advice and input from Paul Mohai and Sandra Arlinghaus. Mohai helped Ard generally (as her PhD advisor) to develop environmental justice connections (Mohai and Saha, 2006). Arlinghaus created various base map files (per personal communications, 2008) from which Ard then imaginatively integrated data over the years along with a variety of environmental justice considerations. The results of Ard's work were submitted to a Google contest and Ard won one of the top worldwide awards (as one of two contest winners), in the student category, for that work (Google Maps, 2009). Figure 9.1 shows an animation derived from that model. To drive around in the virtual world, and to study that model from various perspectives at leisure, download the necessary software and files from the links in Figure 9.1 and study the associated images suggesting how to proceed.

Methods and Timeline Discussion

When the reader has linked to the animation using the QR code above, it's interesting to see change patterns over time, with one replacing another. Figure 9.2 shows an image in which the timeline is clearly visible.

Creating a timeline in Google Earth is not as simple as clicking on a check box, although using it (once created) is that simple. The method and syntax associated with this timeline creation, and other model elements' creation, is shown below.

The first step to loading the geographic data into Google Earth was to load the RSEI data into ArcGIS. The data set for the Detroit Metro Area was pulled out of a larger national database that houses pollution risk scores for every 1×1 km^2 in the United States. The data is categorized by latitude and longitude of the central point of the square, as well as being linked to census block identification numbers. From this set, all squares within the three-county area surrounding Detroit, Michigan were selected. Information about the type and amount of pollution estimated to affect any particular site is associated with

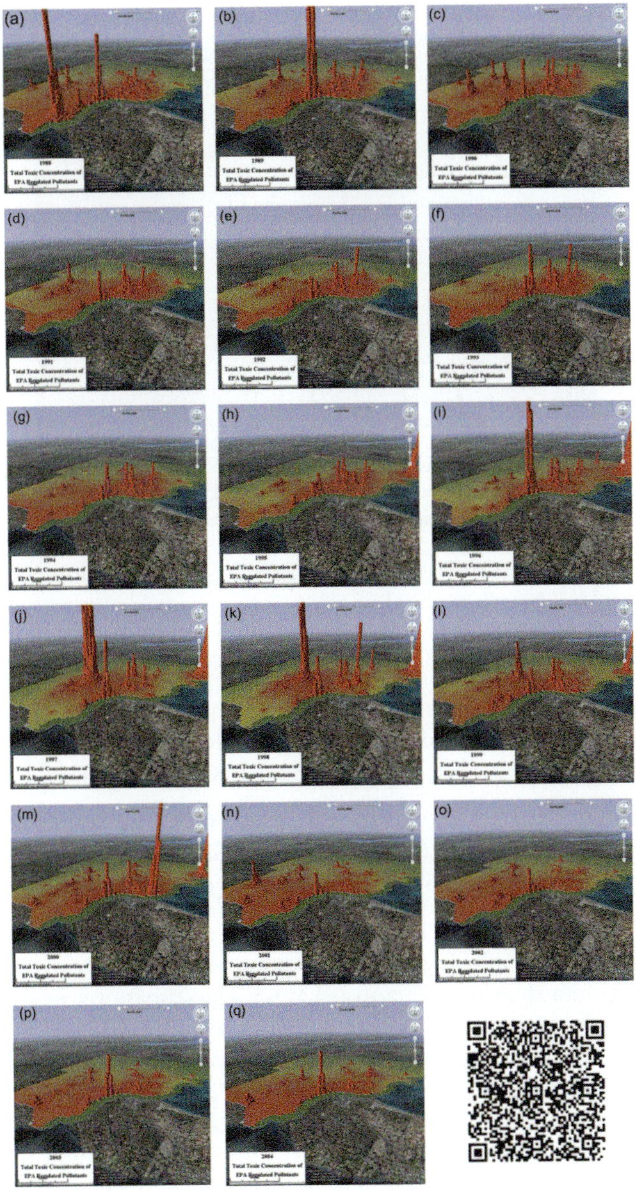

Figure 9.1 *Temporal sequence of perspectives on Ard's model (QR code: http://www-personal.umich.edu/%7Ecopyrght/image/solstice/sum09/AnimodelcroppedArd.gif). (Source: IMaGe, ©2009. http://hdl.handle.net/2027.42/63017. Used with permission.)*

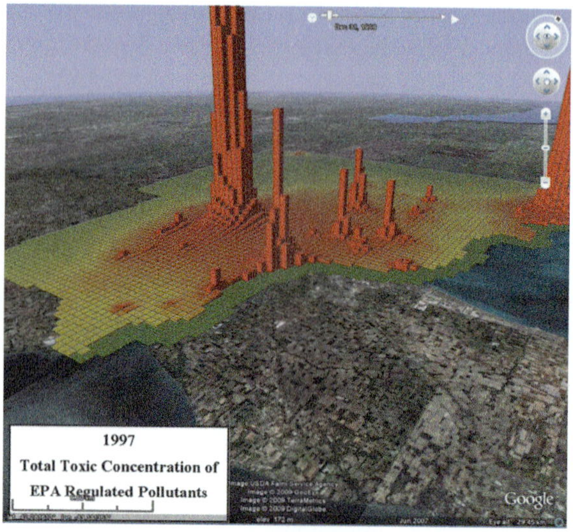

Figure 9.2 *Note the horizontal timeline at the top, just right of center. (Source: IMaGe, ©2009. http://hdl.handle.net/2027.42/63017. Used with permission.)*

the latitude/longitude coordinates. The type and amount of pollution values are used to estimate a total toxic concentration for each square. These toxicity scores are comparable across different areas and time periods.

Once a suitable data layer was open in ArcGIS, the symbology options were opened and choices made concerning the toxic concentration of EPA-regulated pollutants (seen as TOXCON in the example below) as the value field (Figure 9.3).

To determine which symbology colors to use, two questions were considered: (1) How is the data distributed? and (2) What classifications would be useful? Unfortunately, the EPA warns that these data should not be used to determine the chance of illness for the exposed population. Thus, comparison of the toxicity of one area with another seems to be one of only a few natural choices. Because of this limitation, the only value that can be understood through the coloration of the data is where that square is located in the whole distribution of the data. Thus, 'Natural Breaks', which identifies naturally occurring groupings in the data (Jenks, 1967), was selected as the data-partitioning scheme. The hue, saturation, and value (HSV) method was used, which takes a slice across the color wheel. The slice between red and green, typically associated with safe (green) and unsafe (red), was chosen as the color pattern. This choice allows viewers to visually consider which areas are safer than others.

With the map colored, the next step was to export it to Google Earth. For this purpose, the program Export to KML was downloaded (free) and installed. This program installs a tool on the ArcGIS (Esri) toolbar that has an icon as

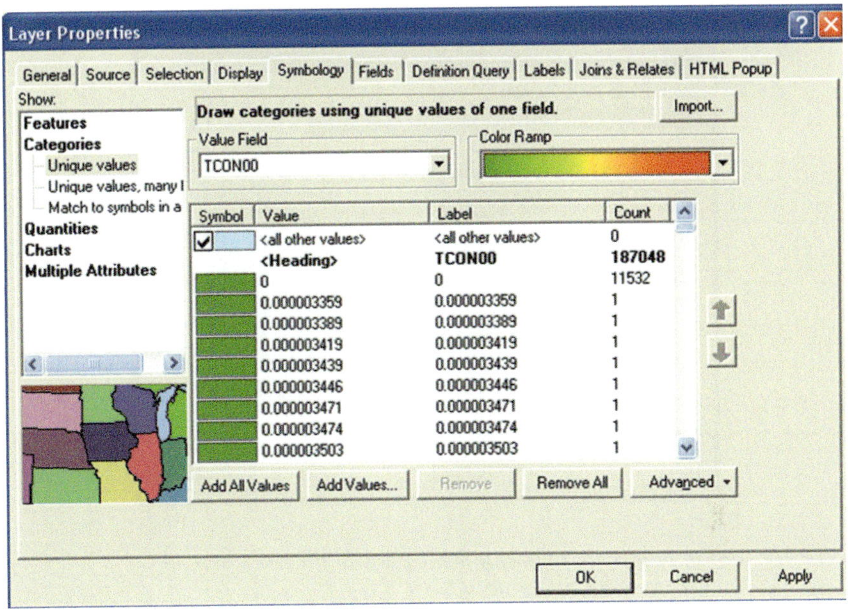

Figure 9.3 *Screen capture from Esri's ArcMAP, Version 9.3.*

shown in Figure 9.4. (Note: those who do not have administrator privileges for ArcGIS can get similar results using standalone software for converting shape files to .kml files. One such standalone program is Shp2kml 2.) The 'Export to KML' tool allows users to upload their shapefiles from ArcGIS to Google Earth. Figure 9.5 shows a screen capture of the option pad that comes up from clicking on the icon in Figure 9.4 (when installed in ArcGIS).

In this option pad, the Toxic Concentration (TOXCON) layer was selected as the layer to export and to use to represent the height (as TOXCON). The data set is highly skewed, with most areas having low levels of toxic concentration and only a few areas having high levels of concentration. Thus, by choosing the height as the actual value of the toxic concentration for cells, users can not only visualize areas of general toxicity through color but can also observe when and where there are *extraordinarily* high values of toxicity.

Figure 9.4 *Export to KML icon that appears on the ArcMAP toolbar.*

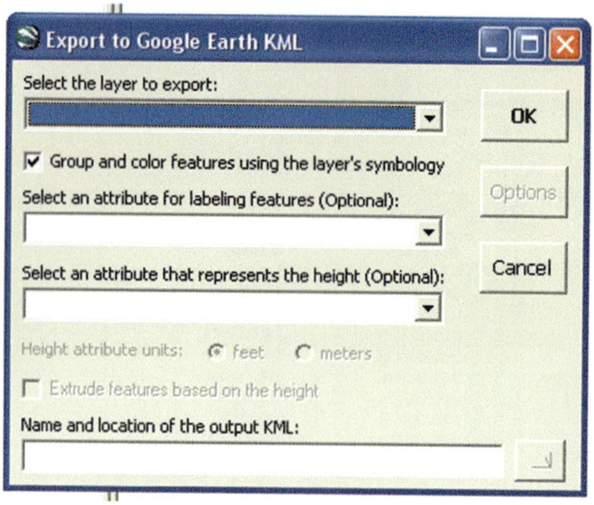

Figure 9.5 *Option pad from Export to KML plug-in for ESRI ArcGIS.*

Thus, viewers can not only clearly visualize actual data patterns but can also be guided toward suggested local regions for added research. Such patterns may be clearly obvious when visualized, although they may not be obvious when comparing columns of numerical data from one spreadsheet to another.

The symbology process was repeated for every year from 1988 to 2004 and each year was uploaded individually to Google Earth (Figure 9.6). Once these layers are in Google Earth they can be joined together under one project. In Figure 9.6, note that each layer uploaded to Google Earth is labeled as TC## (TC is for toxic concentration and ## is for the year), MITRI (which stands for Michigan TRI data), NAD27 (the projection used), and SOUTHMI (South Michigan, the spatial reference).

If one of these layers is expanded in Google Earth by clicking the plus sign button (+) next to the layer, a title key becomes evident. Right-click this layer, choose copy, then paste it into a Notepad document. Upon doing so, the following syntax will appear.

```
<?xml version="1.0" encoding="UTF-8"?>
<kml xmlns="http://www.opengis.net/kml/2.2" xmlns:gx="http://
www.google.com/kml/ext/2.2" xmlns:kml="http://www.opengis.net/
kml/2.2" xmlns:atom="http://www.w3.org/2005/Atom">
<ScreenOverlay>
  <name>TitleKey</name>
  <TimeSpan>
    <begin>2004</begin>
    <end>2005</end>
```

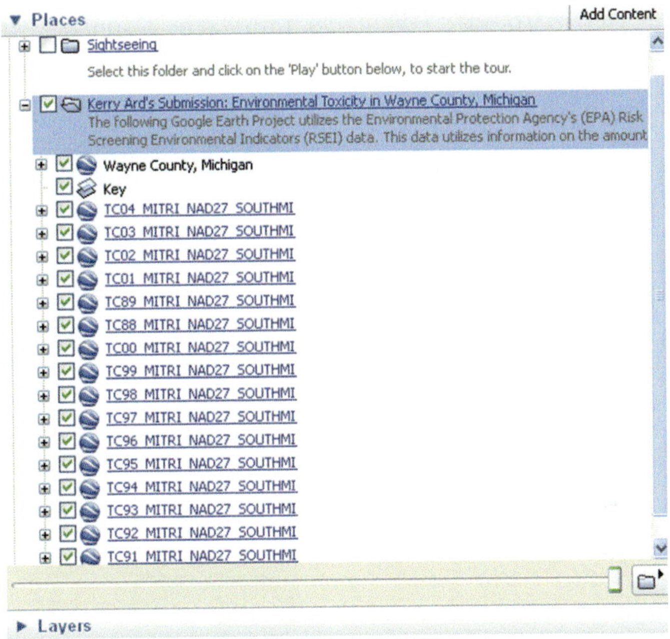

Figure 9.6 *Screen capture showing naming pattern of files for Google Earth.*

```
  </TimeSpan>
  <Icon>
  <href>http://www-personal.umich.edu/~kerryjoy/2004.PNG</href>
  </Icon>
  <overlayXY x="0.02" y="0.1" xunits="fraction" yunits=
   "fraction"/>
  <screenXY x="0.02" y="0.02" xunits="fraction" yunits=
   "fraction"/>
  <rotationXY x="0.5" y="0.5" xunits="fraction" yunits=
   "fraction"/>
  <size x="0" y="0" xunits="fraction" yunits="fraction"/>
</ScreenOverlay>
</kml>
```

Several things are suggested by this syntax. It shows a <TimeSpan> command. This command tells Google Earth to show the icon image (listed as http://www-personal.umich.edu/~kerryjoy/2004.PNG) during the years 2004–2005. The icon image is a simple image created in the program Paint that has the year and the phrase "Total Toxic Concentration of EPA regulated Pollutants." The overlay information below the <Icon> command places this image in the appropriate spot in Google Earth. Each layer of pollution data has been given a title key like the one discussed as well as a 'time span' linking it to an 'icon' image that is suitable for that year.

Similarly, in Figure 9.6 note that under the Wayne County, Michigan layer there is a file called 'Key'. Right-click on this, paste it into Notepad, and the following syntax appears.

```
<?xml version="1.0" encoding="UTF-8"?>
<kml xmlns="http://www.opengis.net/kml/2.2" xmlns:gx="http://
www.google.com/kml/ext/2.2" xmlns:kml="http://www.opengis.net/
kml/2.2" xmlns:atom="http://www.w3.org/2005/Atom">
<ScreenOverlay>
  <name>Key</name>
  <TimeSpan>
    <begin>1988</begin>
    <end>2005</end>
  </TimeSpan>
  <Icon>
  <href>http://www-personal.umich.edu/~kerryjoy/key.PNG</href>
  </Icon>
  <overlayXY x="0.02" y="0.1" xunits="fraction"
   yunits="fraction"/>
  <screenXY x="0.66" y="0.1" xunits="fraction"
   yunits="fraction"/>
  <rotationXY x="0.5" y="0.5" xunits="fraction"
   yunits="fraction"/>
  <size x="0" y="0" xunits="fraction" yunits="fraction"/>
</ScreenOverlay>
</kml>
```

Unlike the syntax discussed previously, we can see that the <TimeSpan> command for this syntax is from 1988 to 2005. This means the icon image that it is linked to will be showing during this entire animation. This image is shown in Figure 9.7 and is a useful guide to have shown for the entire animation.

Each of the toxic concentration layers has a <TimeSpan> command in them. Google Earth will automatically read these and give the option to animate the

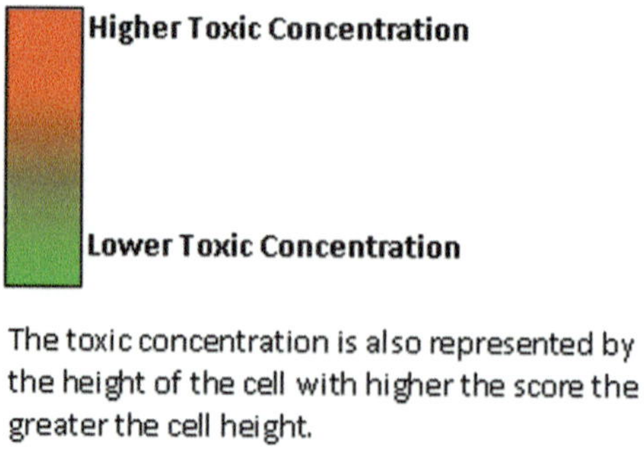

Figure 9.7 *Icon serving as part of a legend for the model of Figure 9.1.*

Figure 9.8 *Time animation toolbar in Google Earth.*

file by showing a toolbar (shown in Figure 9.2 and also, for ease of reference, in Figure 9.8). To view the animation, just click the play button. If needed, change the speed setting by clicking on the wrench image.

Conclusion

This project has helped to visualize a complex data set with multiple inputs for chemicals, concentrations, toxicities, and amounts, all changing over time, as noted in Ard (2009a,b). Two-dimensional representations of such data are limited. Google Earth provides the opportunity to visualize data in 3D and drive through it in order to look at complex spatial patterns from various perspectives (Arlinghaus, 2005). The model presented here does not utilize all options provided by this program. In future projects it might be useful to change the opacity of the layers in order to see what landscape features underlay these areas, such as schools, lakes, parks, and so on (Arlinghaus, 2008). As the capabilities of Google Earth continue to expand, I am confident that it will prove to be a useful tool to educate the public about environmental risks and how their family and communities might be affected (Naud, 2009).

References

Ard, K. 2009a. Air pollution changes in the Detroit Metro Area from 1988–2004. *Solstice: An Electronic Journal of Geography and Mathematics* 20(1).

Ard, K. 2009b. Presentation on GooglEarthDay, April 22, University of Michigan, held in the School of Natural Resources and Environment. http://www-personal.umich.edu/~sarhaus/2009program.pdf.

Arlinghaus, S. 2005. *Spatial Synthesis, Volumes I and II.* Ann Arbor, MI: Institute of Mathematical Geography. http://www.imagenet.org.

Arlinghaus, S. 2008. Personal communications via email with attachments (January 20, July 5, July 18, July 26, September 16, October 10).

City of Portland, Bureau of Planning. 2019. Export to KML plug-in for ESRI ArcGIS. http://arcscripts.esri.com/details.asp?dbid=14273.

EPA's Detailed information about RSEI data.

Esri. 2019. ArcGIS software. http://www.esri.com.

Google. 2009. And the award goes to... (blog). March 9. https://maps.googleblog.com/2009/03/and-award-goes-to.html.

Jenks, G. F. 1967. The data model concept in statistical mapping. *International Yearbook of Cartography* 7: 186–190.

Mohai, P., and R. Saha. 2006. Reassessing racial and socioeconomic disparities in environmental justice research. *Demography* 43(2): 383–399.
Naud, M. 2009. Update on municipal applications of Google Earth. GooglEarthDay presentation, School of Natural Resources and Environment, University of Michigan.
Shape to KML standalone software. Shp2kml 2.

3D Tree Inventory: Geosocial Networking

An Ann Arbor Before-and-After Study

David E. Arlinghaus and Sandra L. Arlinghaus

***Figure 10.0** (a) 'Before' .kmz for download (http://www.mylovedone.com/image/solstice/sum12/Geosocial.kmz). (b) 'After' .kmz for download (http://www.mylovedone.com/image/solstice/win12/TreeInventoryJuly2012.kmz). (c) Spatial word cloud summary, based on word frequency, made using Wordle.*

Geosocial networking is a form of social networking with geographical capabilities (e.g., 'geotagging' or 'geocoding') that compress information and make possible additional social dynamics (Arlinghaus, 2012; Friedemann, 2012). Texted location information or mobile phone tracking are two ways that such social dynamics might be enriched through this form of networking. We illustrate these possibilities in the environmental context of developing a small study that bridged the communication gap between a local municipal project and neighbor concern about that project.

When Washtenaw County, Michigan embarked on a major stream bank erosion control project, neighbors' concerns about the impact on their immediate surroundings became paramount. Indeed, when that project entered heavily forested residential lands adjacent to the slumping stream, environmentally aware residents quite naturally became concerned for the trees and wildlife that would necessarily be destroyed or disturbed. They were far more concerned about the unseen possibilities of tree loss than they were about the clearly visible heavy bank erosion. Thus, it seemed evident that offering them visualization and ongoing capability to monitor the project might allay some of their fears about what they could not see but only imagine. In the spirit of making that communication a two-way street, all the spatial thinking and derived visual evidence was offered not only to neighborhood organizations but also to municipal authorities involved in the project. Such offerings were made during a sequence of meetings: some open and some not so open. The intent was to encourage neighbor support for a necessary but controversial project that, in the long run, would offer substantial environmental enhancement to nearby affected properties as well as to the entire creekshed.

Before …

The county coded its easement with pink flags. It tagged selected large trees or otherwise interesting vegetation with a blue band if they were to be removed; it tagged trees within the easement with a red band if they were to be left alone. All vegetation within the easement, except trees or shrubs carrying red tags, was to be removed. Color was critical; a simple red/blue confusion could cost a tree its life!

We used Google Earth, together with a GPS-enabled smartphone, to make an inventory of trees present, along a half-mile stretch of the creek, before the project began. The first author did all the photography, in an area of about 17 acres, with a smartphone that geotagged the images. He transmitted the set of images to the second author, who did the mapping using a combination of GeoSetter and Google Earth (Figure 10.1).

The accuracy of the geotagging of the photos was limited by several factors. First, the software in the smartphone has limits. Second, the geotagging of the tree is actually the geotagging of the camera position, rather than of the tree position. The photographer attempted to stand at a consistent distance from trees to ensure precision (but that is difficult in a densely wooded area). The level of precision, however, was quite good—trees were in correct relation to each other and in close-to-correct relation to dwelling units.

***Figure 10.1** Pink arrows mark flags showing county drain easements. Red balloons mark trees to be saved within the easement. Blue balloons mark trees to be cut. (Source: IMaGe, ©2012. http://hdl.handle.net/2027.42/91628. Used with permission.)*

The geotagged camera images were downloaded directly to a computer by plugging the smartphone into a desktop computer. All 81 images were stored in a single folder. That folder was then uploaded to the free GeoSetter software. From there, the geotagged images were batch-uploaded to Google Earth in a single operation (rather than entering each one individually). GeoSetter was able to take the underlying geocoded coordinates from the camera images, as well as the images themselves, and make them correspond to the underlying coordinate geometry in Google Earth. We made color decisions to correspond with the actual colors of tags used on vegetation.

Using this strategy, the photo and Google Earth coordinates were guaranteed to be accurately registered. Hand placement would not offer that level of accuracy. Overall, the results were sufficiently precise (although not accurate) to offer local residents a clear picture of what was going to happen in wooded areas. When the camera GPS coordinates were obtained, a photo of the tagged item was also taken. Figure 10.2 shows a photo displayed on the Google Earth surface pointing to the identified red-tagged tree. Figure 10.3 shows a similar configuration of photo in relation to Google Earth base pointing to the identified blue-tagged tree. These pointing associations are all accurate. Download the linked .kmz file, open it in Google Earth, and you will see associations of this sort for all 81 trees marked by the county before the stream bank restoration project was implemented.

Figure 10.2 *Photo mounted in Google Earth. Note that the photo has a pointer on it that points to the correct balloon location. (Source: IMaGe, ©2012. http://hdl.handle.net/2027.42/91628. Used with permission.)*

Figure 10.3 *Blue-tagged tree. (Source: IMaGe, ©2012. http://hdl.handle.net/2027.42/91628. Used with permission.)*

Prior to implementation, the neighborhood association president and the creator of the Google Earth display met with the lead county official and the lead engineer on the project to ensure a cooperative approach to file usage with neighbors who also had access to the same files. Geosocial networking had begun as maps, and archived photographic information stored with the maps was now fixed at a 'before' time point.

After ...

After the removal of trees and the completion of the stream bank stabilization, we tracked the progress of the project. Figures 10.4 through 10.10 show a sequence of screen captures from the field photographs. Figures 10.7 through 10.8 show the successes and failures (related to a drought) of larger, staked plantings. Figures 10.7 and 10.8 show two of the many stream bank photos (from among the 191 new photos added to the 81 'before' photos) designed to illustrate the broader vegetation restoration.

Captions reveal some of what can be noted; however, to get a full view, download the linked .kmz file at the top of this chapter and open it in Google Earth. In that way, the reader of this material can follow along with what happened in this territory without having to walk through the somewhat difficult terrain

Figure 10.4 *A large number of new trees were planted; these plants were larger than shrubs but not huge trees. They needed to be staked but could easily be planted with a shovel. (Source: IMaGe, ©2012. http://hdl.handle.net/2027.42/94573. Used with permission.)*

Figure 10.5 *The staked trees were planted in an area not serviced by the condominium association sprinkler system; this image clearly shows the effects of a drought in early summer. (Source: IMaGe, ©2012. http://hdl.handle.net/2027.42/94573. Used with permission.)*

Figure 10.6 *Some new trees did well, thanks to individual attention from local residents. Others did not. (Source: IMaGe, ©2012. http://hdl.handle.net/2027.42/94573. Used with permission.)*

Figure 10.7 *Distribution of photos of generalized bank vegetation. (Source: IMaGe, ©2012. http://hdl.handle.net/2027.42/94573. Used with permission.)*

Figure 10.8 *The bank here appears to be filling in nicely with a mix of vegetation planted early enough to benefit from rains prior to the drought and to watering by persistent residents during the drought. Other bank locations were more problematic, particularly in association with terrain steepness. (Source: IMaGe, ©2012. http://hdl.handle.net/2027.42/94573. Used with permission.)*

and follow where the county is spending taxpayer dollars on an important environmental restoration project.

In addition to its intended purpose to bridge communication between neighborhood groups and municipal authorities, the geosocial network was helpful in pinpointing which locations to water during the drought. Six years later, in 2018, the new growth is established, and the creek banks now appear stabilized.

References

Arlinghaus, D. E. 2012. Field work, Huron Chase Condominiums, Ann Arbor, MI. With permission of the Condominium Association Board of Directors.

Arlinghaus, D., and Arlinghaus, S. 2012a. Geosocial networking: A case from Ann Arbor, Michigan. *Solstice: An Electronic Journal of Geography and Mathematics* 23(1). Retrieved from http://www.mylovedone.com/image/solstice/sum12/Geosocial.html.

Arlinghaus, D., and Arlinghaus, S. 2012b. Part 2: Geosocial networking: A case from Ann Arbor, Michigan. *Solstice: An Electronic Journal of Geography and Mathematics* 23(2). Retrieved from http://www.mylovedone.com/image/solstice/win12/Geosocial2.html.

Friedemann S. 2012. GeoSetter. http://www.geosetter.de/en (last accessed June 15, 2012).

Washtenaw County. 2012. Mallett's Creek restoration project: Final report. http://www.ewashtenaw.org/government/drain_commissioner/dc_webWaterQuality/malletts_creek/dc_mc_mcrp.html (last accessed June 15, 2012).

11

3D Georeferencing: Downtown Ann Arbor Creek Images, Revisited

Sandra L. Arlinghaus

Figure 11.0 *Spatial word cloud summary, based on word frequency, made using Wordle.*

"Eureka!" Archimedes exclaimed, so the story goes, when he jumped into a full bathtub and water overflowed onto the ground. Water was displaced by his mass and he discovered his 'Principle of Displacement', that the volume of displaced fluid is equal to the volume of the submerged object. As is characteristic of 'principles', 'concepts', and 'abstraction', these are ideas that often have far-flung applications both in obvious ways and in unforeseen ways.

In an urban setting, it is often a matter of debate among planners, engineers, municipal authorities, real estate developers, and members of the public as to whether one should be permitted to build new structures within an existing urban floodplain. If one focuses on Archimedes' principle, it seems obvious that no new construction should ever be permitted within a floodplain.

The floodplain was carved out by an overflowing creek. The floodplain will get filled again, sooner or later, and when it does, where will the water displaced by new construction go? It will add to the volume of water associated with the existing floodplain, forcing some to overflow elsewhere. A 100-year floodplain may overfill in two consecutive years. Some might argue that buildings built on stilts will not displace as much water in a flood as would the same building placed on the ground—true, but others worry about trees washing down the creek from upstream and getting caught on the building stilts and damming the creek, much like a beaver's dam. Still others claim that offsite retention ponds can capture extra flow from upstream in the creekshed, leaving room for the extra volume of water coming from a large new building within the floodplain. All of the arguments on the latter side of this equation have one thing in common: a fundamental desire to convert floodplain parcels into money-makers for the municipality and its business interests by constructing large buildings in them. An equation with physics concerns on one side of the equation and economics concerns on the other side can be a difficult one to balance. There are intricate means already available to assess the appropriate physical and economic concerns; getting the physics and economics folks to negotiate an agreement acceptable to both, within interpretations of local municipal laws and practice, can be a daunting task.

The task can become even more difficult when the general public gets involved, as it generally does during the formal, legal planning process. Simple visuals can go a long way in teaching the public, and others, to understand the elementary components of the issues involved.

Georeferencing in Google Earth

Allen Creek runs along the west side of the downtown of Ann Arbor, Michigan. The virtual reality model of Chapter 6, coordinated with appropriate music and other sound effects, along with a hazmat linkage, went a long way in providing the public and others with a display concerning the possible flooding of Allen Creek and its effects on the downtown that was convincing enough to pique interest. Those displays, however, were not georeferenced to the Earth and were not easily created or maintained by the public, nor were they easy to change over time.

Here, we employ a simple method, based on using Google Earth and Trimble/Google SketchUp to present visuals that members of the public can recreate on their home computers using only free software. Because they can recreate the images at any given time, the public can track change over time. Citizen science opportunities are enhanced.

One simple way to consider where a flood might go is to think of filling the terrain with water to a certain contour level—rather like filling a cup to overflow into its accompanying saucer, then table, then floor, and beyond. Google Earth, supported by SketchUp, permits such visualization when coupled with a concept from multivariable calculus. Google Earth is easy to use, and it is free; CityEngine is another tool that is enabling planners with 3D modeling techniques; many others will no doubt be forthcoming. Google Earth has the capability to display terrain (by clicking on the 'Terrain' checkbox, often near the bottom left side of the screen, although this position could vary with the software version). It does not display contours, although it clearly has digital information within it to recognize terrain height.

Contours are lines that are probably familiar to most readers, displayed on topographic maps, for example, as lines of equal elevation above a base level. Their theoretical origin is the level curve of a surface: that is, given a surface $z = f(x, y)$, intersect a plane, $z = k$ (parallel to the xy plane or base level), with that surface. The line of intersection traced out on the surface is a contour at elevation k.

So, in order to determine contour lines in Google Earth (within a local area where the influence of the curvature of the Earth is insignificant), it will suffice to insert a plane (partial) parallel to the base level at desired elevations. Then, color the plane a translucent blue color (like water), and all parts of the Google Earth terrain less than the selected value will be blue (covered with water) and all parts above will remain visible as terrain. It is a simple 'greater than or less than' technique, with equality on the blue plane. Any given point in the area under consideration—that is, not on the plane—either lies above the plane or below the plane.

Figure 11.1 shows the effect of introducing the blue plane at an elevation of 860 ft. into the downtown of Ann Arbor in Google Earth. Turn on the 3D buildings layer. Rotate the file, take a closer look by magnifying the file, translate it and move it around, and combine these various rigid motions of navigation. Figures such as these are useful in public demonstrations. A natural question that arises from many in the audience is, "Where is this water going in relation to my home or office?" Anyone who has ever taught mapping knows that this sort of question is a typical first question. It is a good idea to have a stockpile of ready examples.

Beyond the demonstration/education value, there is the general value of tracking change over time. Google does most of the work for us, as its experts insert new buildings and new changes in the terrain. These changes will, of course, also show up in new versions of software and they may or may not continue to work well with the plane areas, calculated once only for systematic comparison purposes (as long as the base level, WGS84, within Google Earth does not change). If they do, that's great, and the elegance of the process is

Figure 11.1 *Contour (860 ft.) used to create a flood in downtown Ann Arbor, MI. (a) Overview. (b) Closer look at Main Street. View in 2018 Google Earth. (Source: IMaGe, ©2018 for new creation using older files. Used with permission.)*

intellectually appealing. However, if they do not, it is a relatively straightforward manner to create a new plane and use that. The next section gives a general idea of the process; it remains 'general' to focus on concepts that may endure independent of software changes (e.g., the Microsoft release in 2018 of all of the buildings in the United States into the public domain).

Software Usage: General Review

Google Earth is straightforward.

Google/Trimble SketchUp.

- Launch New File.
- Geo-locate the image.
  - File | Geo-location | Add location.
  - Navigate to location on map. Select region. Grab.
- Preparation for use of geo-located image.
  - Select image. Left-click on it so it is outlined as a red plane.
  - Right-click on selected image. Choose 'Unlock'. Now the image is outlined as a blue plane.
- Add terrain.
  - File | Geo-location | Show terrain. Click on the latter.
  - Now the image is wrinkled by the terrain.
  - Select image. Left-click on it so it is outlined as a red box. The height of the red box shows relative relief.
  - Right-click on selected image. Choose 'Unlock'. Now the image is outlined as a blue box.
  - Right-click on image. Choose 'Explode'. Now the entire wrinkled plane is hatched in a blue pattern.
- Add color: Use 'Colors and mirrors' to select a translucent glass tinted blue (or whatever you want, but choose translucent if you want to be able to see through the flood).
- Raise the image to the desired height; this may involve some experimental work. Hold the 'Shift' key down to keep the motion in relation to the blue axis (z-axis).
- Hide the underlying terrain image so that only the elevated plane shows.
- Export the plane as a .kmz file.
- Load into Google Earth. (If the position looks wrong, go back and tinker with the elevation in SketchUp; there may be variations in technique from one version of software to another.)

References

Arlinghaus, S. L. 2005. Update on the 3D atlas of Ann Arbor: Archimedes in Ann Arbor? *Solstice: An Electronic Journal of Geography and Mathematics* 16(2).

Arlinghaus, S. L. et al. 2006. *3D Atlas of Ann Arbor*, 2nd Edition. Ann Arbor, MI: Institute of Mathematical Geography (http://www.imagenet.org).

IV

GEOMATs, 2000s

This section introduces the GEOMAT (Geographic Events Ordering Maps, Archives, and Timelines) tool as a method for integrating space and time in spatial thinking. The first chapter in this section considers the Varroa mite in this context, again for comparing and contrasting with the methods in the two previous sections. It is followed by other examples covering a wide range of applications and concludes with a conceptual summary.

12

Varroa GEOMAT: Honeybee Mite

Sandra L. Arlinghaus and Diana Sammataro

Figure 12.0 Spatial word cloud summary, based on word frequency, made using Wordle.

Compression of Information Using GEOMAT

Tracking spatial and temporal information on bee pests, such as the Varroa mite, is of equal importance in developed and developing nations. These insect pests know no national boundaries; software and technological capabilities do, however. Thus, we choose as a primary tracking tactic to use simple software that has been globally available for a while, is free to download, and does not require extensive use of online materials and the associated capability for lengthy connection to the internet (Arlinghaus, Larimore, and Haug, 2016; Larimore, Arlinghaus, and Haug, 2005; 2007; 2008). The GEOMAT in Figure 12.1 shows a calendrical timeline for the entire time, a year at a time, together with brief comments from the field study archive. The following timeline is rooted

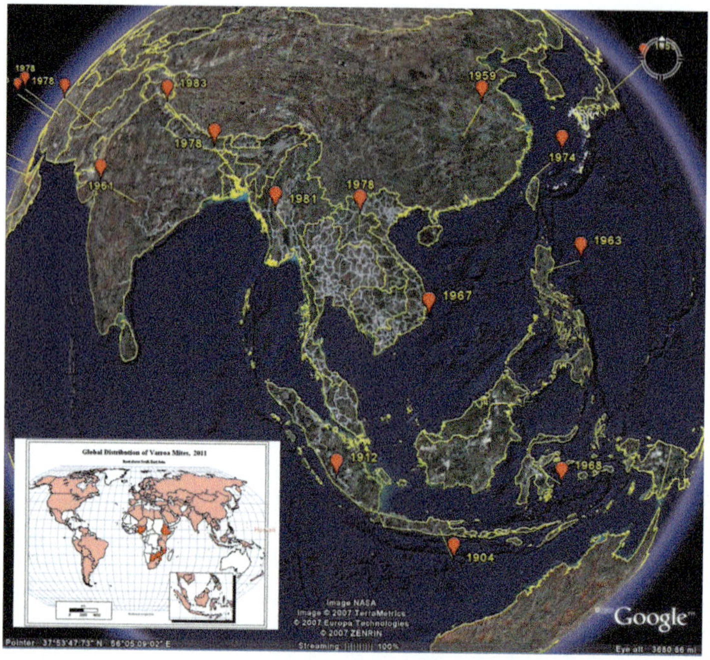

Figure 12.1 *The GEOMAT displays a calendrical timeline rising from a map to display events, and gaps between events, simultaneously. (Source: IMaGe, ©2016. http://hdl.handle.net/2027.42/134732. Used with permission.)*

in a map that combines the images from the animated approach to looking at the Varroa information with the 3D map of that information. The material is all embedded in a 'Table' in Microsoft Windows. An alternate strategy might be to create a table in html. Either approach is effective and easy.

1904	Initial recorded sighting of Varroa on the island of Java, Indonesia.
1905	No new sightings recorded.
1906	No new sightings recorded.
1907	No new sightings recorded.
1908	No new sightings recorded.
1909	No new sightings recorded.
1910	No new sightings recorded.
1911	No new sightings recorded.
1912	Next recorded sighting, on nearby Sumatra island, Indonesia.
1913	No new sightings recorded.
1914	No new sightings recorded.
1915	No new sightings recorded.
1916	No new sightings recorded.

(*Continued*)

1917	No new sightings recorded.
1918	No new sightings recorded.
1919	No new sightings recorded.
1920	No new sightings recorded.
1921	No new sightings recorded.
1922	No new sightings recorded.
1923	No new sightings recorded.
1924	No new sightings recorded.
1925	No new sightings recorded.
1926	No new sightings recorded.
1927	No new sightings recorded.
1928	No new sightings recorded.
1929	No new sightings recorded.
1930	No new sightings recorded.
1931	No new sightings recorded.
1932	No new sightings recorded.
1933	No new sightings recorded.
1934	No new sightings recorded.
1935	No new sightings recorded.
1936	No new sightings recorded.
1937	No new sightings recorded.
1938	No new sightings recorded.
1939	No new sightings recorded.
1940	No new sightings recorded.
1941	No new sightings recorded.
1942	No new sightings recorded.
1943	No new sightings recorded.
1944	No new sightings recorded.
1945	No new sightings recorded.
1946	No new sightings recorded.
1947	No new sightings recorded.
1948	No new sightings recorded.
1949	U.S.S.R.
1950	No new sightings recorded.
1951	No new sightings recorded.
1952	No new sightings recorded.
1953	No new sightings recorded.
1954	No new sightings recorded.
1955	Japan.
1956	No new sightings recorded.

(*Continued*)

1957	No new sightings recorded.
1958	No new sightings recorded.
1959	China.
1960	No new sightings recorded.
1961	India.
1962	No new sightings recorded.
1963	Philippines.
1964	No new sightings recorded.
1965	No new sightings recorded.
1966	No new sightings recorded.
1967	Vietnam.
1968	Indonesia (remainder), Cambodia, Thailand, South Korea.
1969	No new sightings recorded.
1970	No new sightings recorded.
1971	Germany, Paraguay.
1972	Brazil.
1973	No new sightings recorded.
1974	No new sightings recorded.
1975	Argentina, Tunisia, Poland, Romania.
1976	Uruguay, Libya, Finland.
1977	Montenegro.
1978	Pakistan, Iran, Turkey, Czechoslovakia, Laos.
1979	Italy.
1980	Bolivia.
1981	Algeria, Burma.
1982	France, Austria.
1983	Nepal.
1984	Switzerland.
1985	Peru, Spain.
1986	Chile, Malaysia.
1987	United States (conterminous), Canada, Saudi Arabia, Sweden.
1988	Iraq, Bahrain.
1989	Portugal, Morocco, Egypt.
1990	Syria.
1991	Lebanon, Niger, Venezuela.
1992	Mexico, Mongolia, United Kingdom, Alaska.
1993	Grenada, Norway.
1994	No new sightings recorded.
1995	Cuba and Trinidad.
1996	No new sightings recorded.

(*Continued*)

1997	South Africa.
1998	Dominica, St. Lucia.
1999	No new sightings recorded.
2000	New Zealand (north island), Panama, St. Kitts, Tobago, Nevis.
2001	No new sightings recorded.
2002	No new sightings recorded.
2003	No new sightings recorded.
2004	No new sightings recorded.
2005	No new sightings recorded.
2006	New Zealand (south island).
2007	Hawaii.
2008	No new sightings recorded.
2009	Kenya, Uganda, Tanzania.
2010	No new sightings recorded.
2011	Nigeria, Ethiopia, Mozambique, Botswana, Zimbabwe.
2012	No new sightings recorded.

Time Gaps: GEOMAT Archiving

The GEOMAT timeline displays the time gaps more clearly than does the animated map. The animated map displays the full globe at once but does not display layers other than static layers inserted at the outset. The Google Earth globe is dynamic as it permits the insertion of layers in relation to given mapped data, but it cannot display the full globe simultaneously. Thus, the GEOMAT draws in both the animated map and the Google Earth globe as a sort of a base from which the timeline rises as one way of displaying all information, including time gaps. The time gaps highlight opportunities to guide further research.

An important benefit of the GEOMAT involves its ability to serve as an archive for materials so that a researcher can see, at a glance, not only a list of references but also how the research articles are clustered in time. Gaps should once again raise questions: Did the researcher simply fail to find relevant references? Did political decisions of some sort curtail funding that affected research activity? Gaps suggest extra research opportunities.

In Figure 12.2, we illustrate the idea of employing the GEOMAT format in this manner. As the Varroa mite has spread, Sammataro has sent new data, on a regular basis, for additional mapping by Arlinghaus. For many of the updates, the latter simply used a painting package to insert new frames in the animation in order to keep the visual appearance consistent from 1998 forward. For other reasons, she used 3D animation as it became available. There are merits and drawbacks to each.

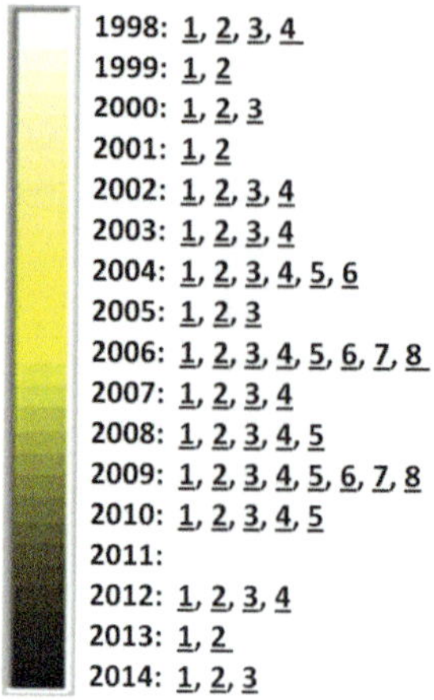

Figure 12.2 *GEOMAT archiving of materials. Pattern of animated maps appearing in, or linked to, Solstice (1998–2014). (Source: IMaGe, ©2016. http://hdl.handle.net/2027.42/134733. Used with permission.)*

It is far easier to comprehend the global diffusion of the Varroa mite looking at the simple Robinson projection animation than it is looking at one side of the beautiful 3D globe display. On the other hand, the globe display is eye-catching, shows good realistic detail, and works well for displaying regional studies in association with terrain. What the GEOMAT can do is incorporate all of these, and more, as the archive of materials in Figure 12.2 suggest.

Future Direction

The mission of this chapter was multifold. First, it was to compare and contrast different mapping tools using the same environmental context: animated maps, Google Earth maps, and GEOMAT, using them on issues associated with the global honeybee population. The process associated with creating a GEOMAT was displayed in some detail for the reader. Second, it was to illustrate the importance of interactivity in being able to combine archives of varying scales and formats and to do so in a way that makes the updating of research a dynamic process. The rest of the section will show uses of the GEOMAT method, perhaps coupled (as was the case here) with other earlier tools.

References

Arlinghaus, S. L., A. E. Larimore, and R. Haug. 2016. Maps, archives, and timelines: Revisiting the 'GEOMAT' project. *Solstice: An Electronic Journal of Geography and Mathematics* 27(2).

Larimore, A. (with S. Arlinghaus and R. Haug). GEOMAT: Geographic Events Ordering: Maps, Archives, Timelines: An historical-geographical method for aiding conflict resolution. University of Michigan course, 2005, 2007, 2008.

Sammataro, D., and S. L. Arlinghaus. 2016. Small hive beetle animaps: Focus on Hawaii. *Solstice: An Electronic Journal of Geography and Mathematics* 27(1).

Groundwater GEOMAT: Google Earth Applications in a Community Information System

Roger Rayle

Figure 13.0 Spatial word cloud summary, based on word frequency, made using Wordle.

Water quality issues often are not uppermost in the minds of homeowners or decision-makers. The water is there, and we tend to take for granted that it is clean, safe, and plentiful. Community groups can work to heighten awareness of these issues (e.g., Scio Residents for Safe Water). New visualization technologies can be very helpful in capturing the attention of groups that might otherwise remain complacent.

One contaminant that may be present in urban water supplies, offering a potential threat to public health, is 1,4-dioxane. It is a substance that may be released from industrial plants, detergents, and other hosts. Once in the system, it is highly mobile, stable, and difficult to remove. Remediation involves substantial commitment from local and federal agencies alike.

What is shown here is a GEOMAT representation of an over 30-year struggle to encourage the appropriate authorities in the State of Michigan to take seriously, and react appropriately to, a dioxane contamination on the west side of Ann Arbor, Michigan.

Static screen shots from Google Earth, from 2006, offer a snapshot view of the situation, georeferenced so that well contamination is visible at a glance (Figures 13.1 and 13.2).

To understand how this situation came about prior to 2006, and to understand where it has gone since 2006, the GEOMAT display georeferences events in time, along a timeline, and roots that timeline in a map showing the contemporary state of events (Figure 13.3).

This periodically updated mashup (Figure 13.3 shows a static image from 2017) is used as a presentation tool for public meetings and as a decision support tool at technical meetings where the state, local government, and citizen representatives review cleanup proposals and make recommendations. In that regard, it appears to be an outstanding free tool to allow community stakeholders to present multifaceted views of reality in a concise, unified format

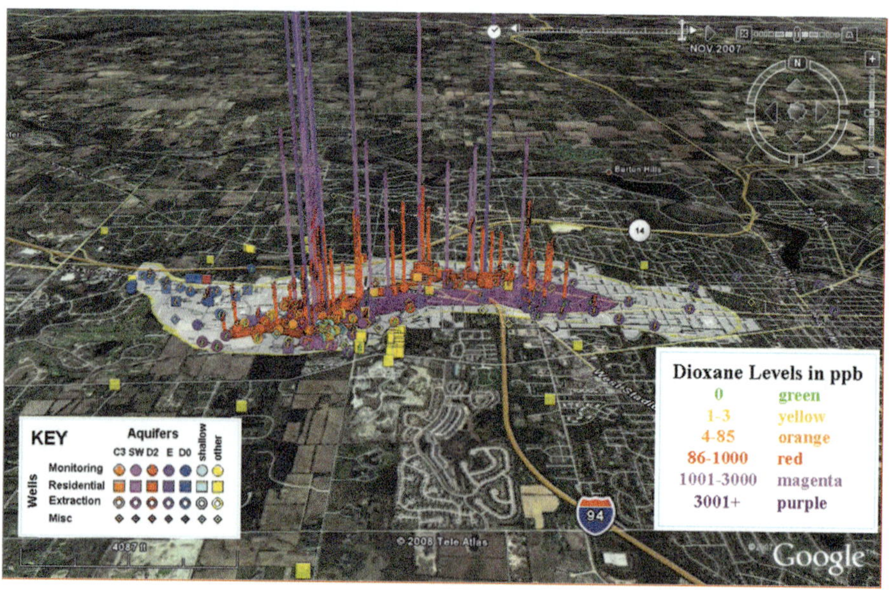

Figure 13.1 *Three-dimensional map with georeferenced bars reflecting various aspects related to contamination issues in the aquifer (Rayle, 2006). (Source: DEQ_PLS1206.mdb from MDEQ (Gelman site sampling and well log data). Recent data available at DEQ–Gelman Access Database. Licensed under a Creative Commons Attribution 3.0 License (http://creativecommons.org/licenses/by/3.0) by Roger Rayle & Scio Residents for Safe Water.)*

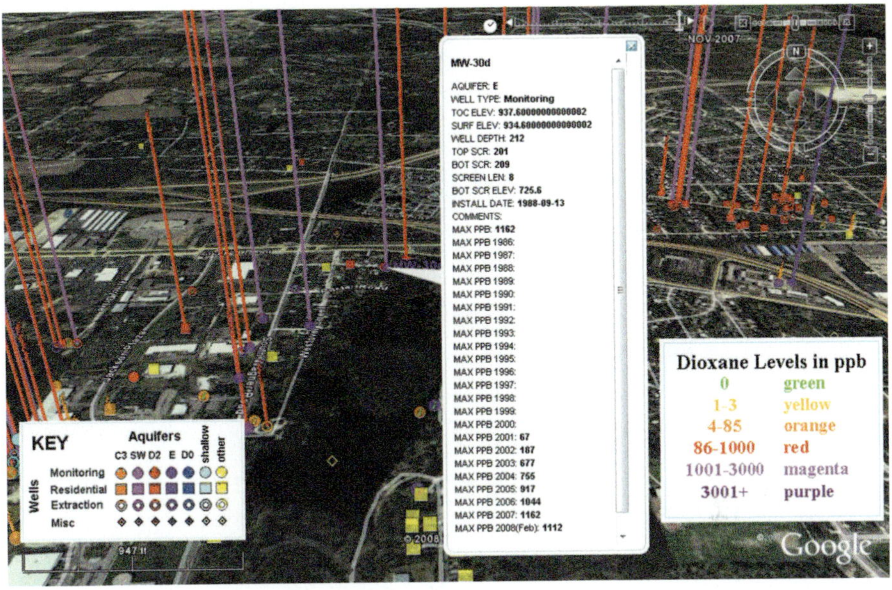

Figure 13.2 *Three-dimensional map with georeferenced bars reflecting various aspects related to contamination issues in the aquifer—a closer look (Rayle, 2006). (Source: DEQ_PLS1206.mdb from MDEQ (Gelman site sampling and well log data). Recent data available at DEQ–Gelman Access Database. Licensed under a Creative Commons Attribution 3.0 License (http://creativecommons.org/licenses/by/3.0) by Roger Rayle & Scio Residents for Safe Water.)*

that can effectively influence decision-makers. A QR code links the reader here to a .kmz file to load into Google Earth so that he/she may navigate the many layers associated with this massive field data accumulation effort (Figure 13.3).

The map in Figure 13.3 indicates that the process created more than a decade ago is one that is still viable and one that can be reasonably updated as times and software interests change. Decisions made at the outset can influence not only the cartographic capability to move forward but also, and perhaps more importantly, can continue to be used to track information to share with the public and influence public policy. Efforts made in this regard over the past 30 years have paid off handsomely for the City of Ann Arbor and its residents.

Gelman begins operation at 600 S. Wagner Rd.	1963	
First seepage lagoon constructed.	1964	
	1965	
Begins using dioxane.	1966	Dioxane used/discharged.

(*Continued*)

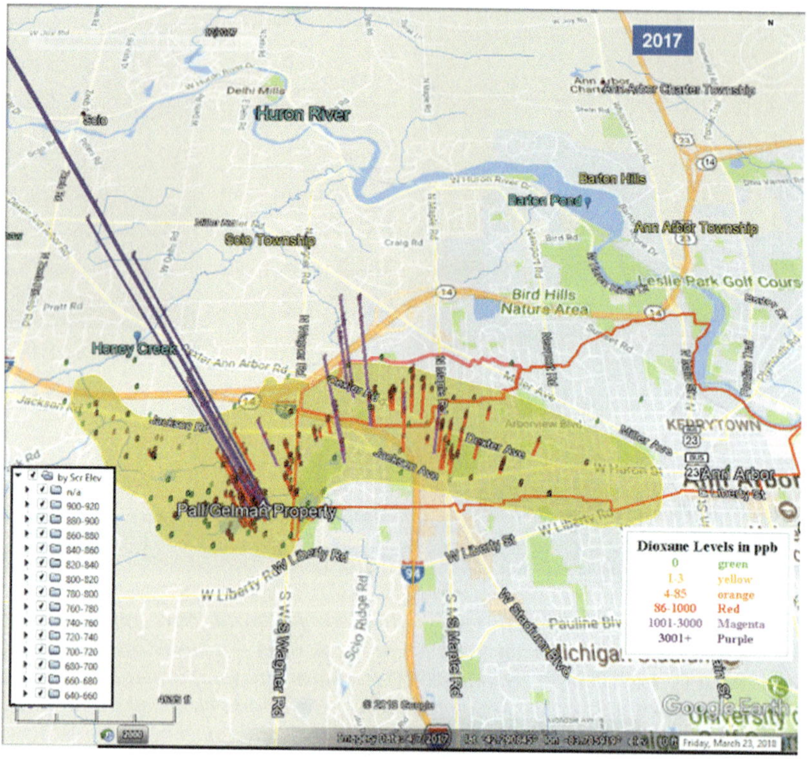

Figure 13.3 *Source for GEOMAT data, Scio Residents for Safe Water. Link to .kmz file to load into Google Earth (Rayle, 2011): http://www.mylovedone.com/image/solstice/sum11/20110314Mashup-NEview.kmz Persistent link: http://deepblue.lib.umich.edu/handle/2027.42/58219. (Source: Map image licensed under a Creative Commons Attribution 3.0 License (http://creativecommons.org/licenses/by/3.0) by Roger Rayle & Scio Residents for Safe Water.)*

Second seepage lagoon constructed.	1967	Dioxane used/discharged.
	1968	Dioxane used/discharged.
	1969	Dioxane used/discharged.
	1970	Dioxane used/discharged.
	1971	Dioxane used/discharged.

(*Continued*)

	1972	Dioxane used/discharged.
3 million gallon lined lagoon constructed	1973	Dioxane used/discharged.
	1974	Dioxane used/discharged.
	1975	Dioxane used/discharged.
Spray irrigation begins	1976	Dioxane used/discharged.
	1977	Dioxane used/discharged.
	1978	Dioxane used/discharged.
	1979	Dioxane used/discharged.
	1980	Dioxane used/discharged.
Deep well constructed and used for waste water disposal.	1981	Dioxane used/discharged.
	1982	Dioxane used/discharged.
	1983	Dioxane used/discharged.
1,4 -dioxane discovered in Third Sister Lake by UM graduate student.	1984	Dioxane used/discharged.
Spray irrigation ceased; first dioxane in wells; use of dioxane ceased.	1985	Dioxane used/discharged.
Spray irrigation ceased; first dioxane in wells; use of dioxane ceased.	1986	Dioxane used/discharged.
Redskin purge to deep well; waste water to AA treatment plant.	1987	Pumped/dumped to mile deep well.
MDNR lawsuit for cleanup filed.	1988	Pumped/dumped to mile deep well.
Mass Balance, 64,000 lbs left?	1989	Pumped/dumped to mile deep well.
	1990	Pumped/dumped to mile deep well.
	1991	Pumped/dumped to mile deep well.
	1992	Pumped/dumped to mile deep well.
Consent judgment entered; Evergreen purge begins.	1993	Pumped/dumped to mile deep well; Evergreen purge.
Purge via Redskin ceased (deep well plugged); Marshy Pilot Test.	1994	Pumped/dumped to mile deep well; Evergreen purge.
Cleanup std chgs from 3 ppb to 77 ppb.	1995	Evergreen purge.
	1996	Evergreen purge.
Pall acquires Gelman Sciences; Core Purge begins.	1997	Evergreen purge; Core treatment/discharge.
	1998	Evergreen purge; Core treatment/discharge.
	1999	Evergreen purge, approx. 600 lbs removed; Core treat/discharge
Evergreen pipeline to Core opens; Cleanup std chgs from 77 ppb to 85 ppb; Unit E contamination 'rediscovered'.	2000	Core treatment/discharge (+ reinjection at Maple Rd); 80,000 lbs left? Judge sets 5-year deadline for cleanup (85 ppb by July 2005).
	2001	Core treatment/discharge (+ reinjection at Maple Rd); Judge sets 5-year deadline for cleanup (85 ppb by July 2005).

(Continued)

Unit E purge begins.	2002	Core treatment/discharge (+ reinjection at Maple Rd); Judge sets 5-year deadline for cleanup (85 ppb by July 2005).
	2003	Core treatment/discharge (+ reinjection at Maple Rd); Judge sets 5-year deadline for cleanup (85 ppb by July 2005).
	2004	Core treatment/discharge (+ reinjection at Maple Rd); Judge sets 5-year deadline for cleanup (85 ppb by July 2005).
Switch from UV/02 to Oc/02 treatment at core; Original PZ created.	2005	Core treatment/discharge (+ reinjection at Maple Rd); Judge sets 5-year deadline for cleanup (85 ppb by July 2005).
CARD formed as permanent replacement for IPC; Maple Rd (Purge/Reinjection begins.	2006	Core treatment/discharge (+ reinjection at Maple Rd).
Pall to sell buildings after moving out production unit.	2007	Core treatment/discharge (+ reinjection at Maple Rd).
	2008	Core treatment/discharge (+ reinjection at Maple Rd).
	2009	Core treatment/discharge (+ reinjection at Maple Rd).
Data errors in Pall's new database; EPA tightens dioxane stds.	2010	Core treatment/discharge (+ reinjection at Maple Rd).
PZ expanded; Pall cuts off DEQ access to online database.	2011	Core treatment/discharge (+ reinjection at Maple Rd).
	2012	Core treatment/discharge; Treatment discharge; DEQ misses deadline to meet stricter EPA dioxane standards 4 years in a row.
	2013	Treatment discharge; DEQ misses deadline to meet stricter EPA dioxane standards 4 years. in a row.
DEQ settles for $500,000 from Pall out of $5 million.	2014	Treatment discharge; DEQ misses deadline to meet stricter EPA dioxane standards 4 years in a row.
Danaher buys Pall.	2015	Treatment discharge; DEQ misses deadline to meet stricter EPA dioxane standards 4 years in a row.
DEQ seeks $700,000 from taxpayers for future monitoring; AA Twp, Scio and Sierra Club ask for EPA enforcement; Judge allows City, HRWC to join case; DEQ issues temporary emergency rule 7.2 ppb dioxane.	2016	Treatment discharge.
DEQ makes 7.2 ppb rule permanent; Site meets EPA Superfund criteria; PA report filed.	2017	Treatment discharge.

Summary of Technique

Before Google Earth came along, the author spent tens of hours every few months creating 2D depictions of new well-sampling data for a local groundwater cleanup, which he has been monitoring as a citizen volunteer since 1993. Now with Google Earth, in several hours, he can generate annual 2D, 3D and 4D mashups showing dioxane plume locations, wells sampled by type and aquifer, surface water features, remediation attempts, and the locations of over 20,000 pollution samples taken since 1986. A bar whose height represents the concentration of the contaminant is shown at the exact X-Y longitude/latitude for each sample location, with the fourth dimension being date sampled (Figure 13.4).

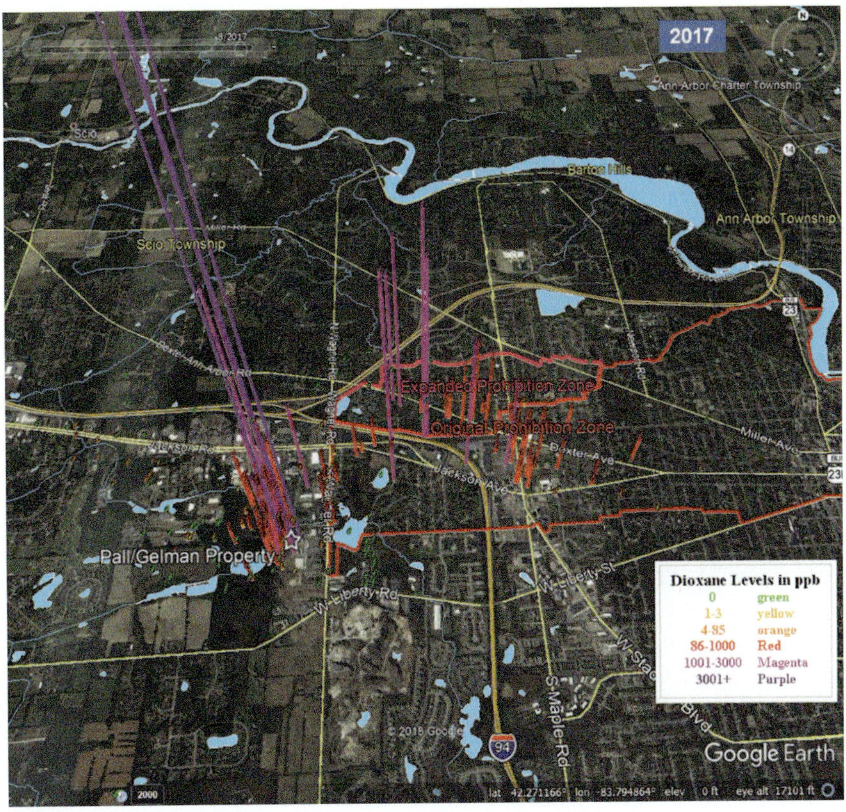

Figure 13.4 *Dioxane screenshot. (Source: DEQ–Gelman database (2017). Licensed under a Creative Commons Attribution 3.0 License (http://creativecommons.org/licenses/by/3.0) by Roger Rayle & Scio Residents for Safe Water.)*

The result viewed on Google Earth gives a fairly clear indication of which ways the contamination plumes are moving, how fast, and at what concentrations. Rayle developed his technique to plot large data sets to Google Earth beginning in April 2007. Working from an initial .kml sample file provided by Arlinghaus, he first constructed a template with the desired colors, line weights, and icons for the categories of data to be plotted. Then he used a simple mail-merge process to generate the placemark kml code from the placemarks section of the sample template. Finally, he copied and pasted the result into the original template, replacing the sample placemarks, and opening the result in Google Earth. He happened to use Word and Excel for the mail-merge and Notepad++ for editing the kml template and final code, but other such programs should work just as well.

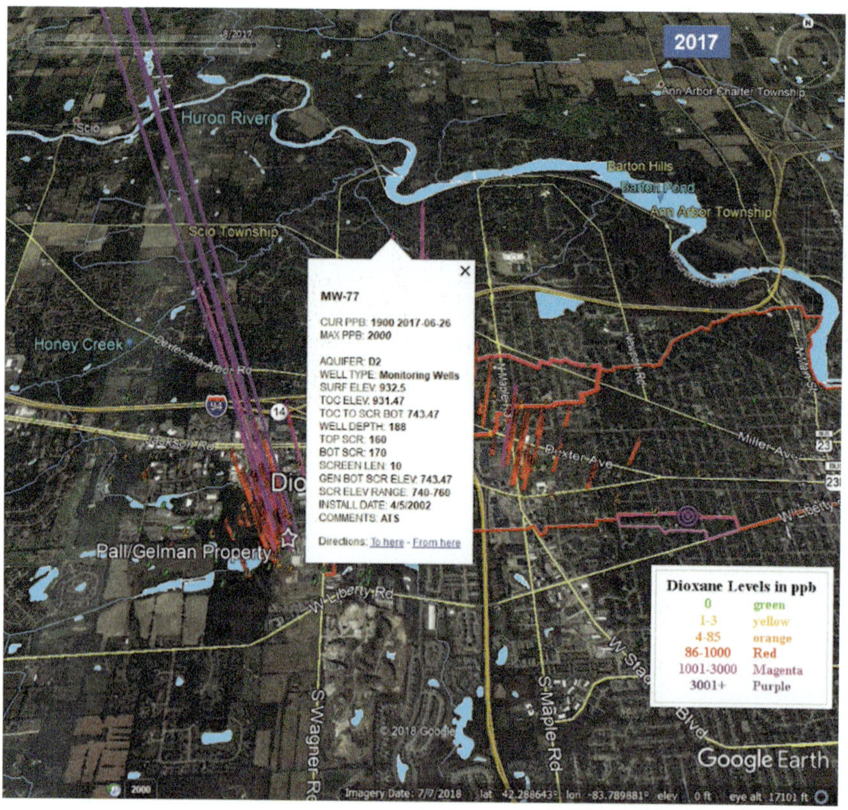

Figure 13.5 *PPB pop-up sample. (Source: DEQ–Gelman database (2017). Licensed under a Creative Commons Attribution 3.0 License (http://creativecommons.org/licenses/by/3.0) by Roger Rayle & Scio Residents for Safe Water.)*

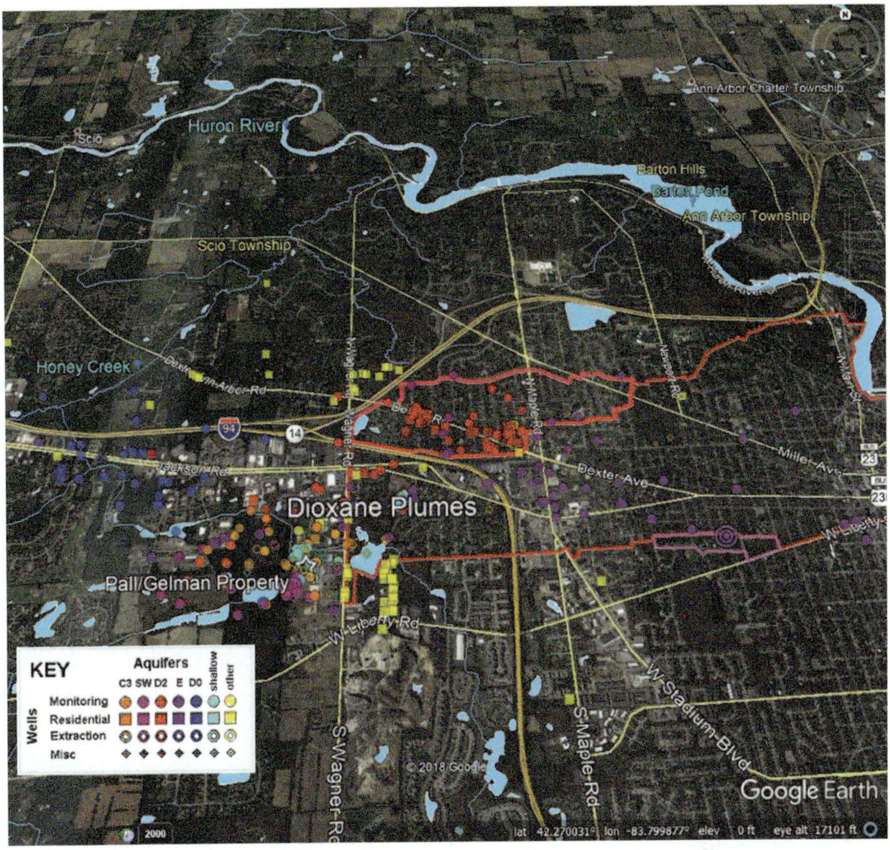

Figure 13.6 *Wells. (Source: DEQ–Gelman database (2017). Licensed under a Creative Commons Attribution 3.0 License (http://creativecommons.org/licenses/by/3.0) by Roger Rayle & Scio Residents for Safe Water.)*

Besides showing the data points as bar graphs, the author has tweaked his templates to show the sample point name and date when one mouses over each placemark and to show a pop-up box of associated well log and dioxane data when one clicks on a sample placemark (Figure 13.5).

Another informative layer shows a surface marker for each of several hundred sample locations color-coded by aquifer and shape-coded by well type (Figure 13.6).

There are two options to show the well-sampling data when one clicks on a well marker. One shows the maximum dioxane ppb per year as a real-time generated graph when a well icon is clicked (Figure 13.7).

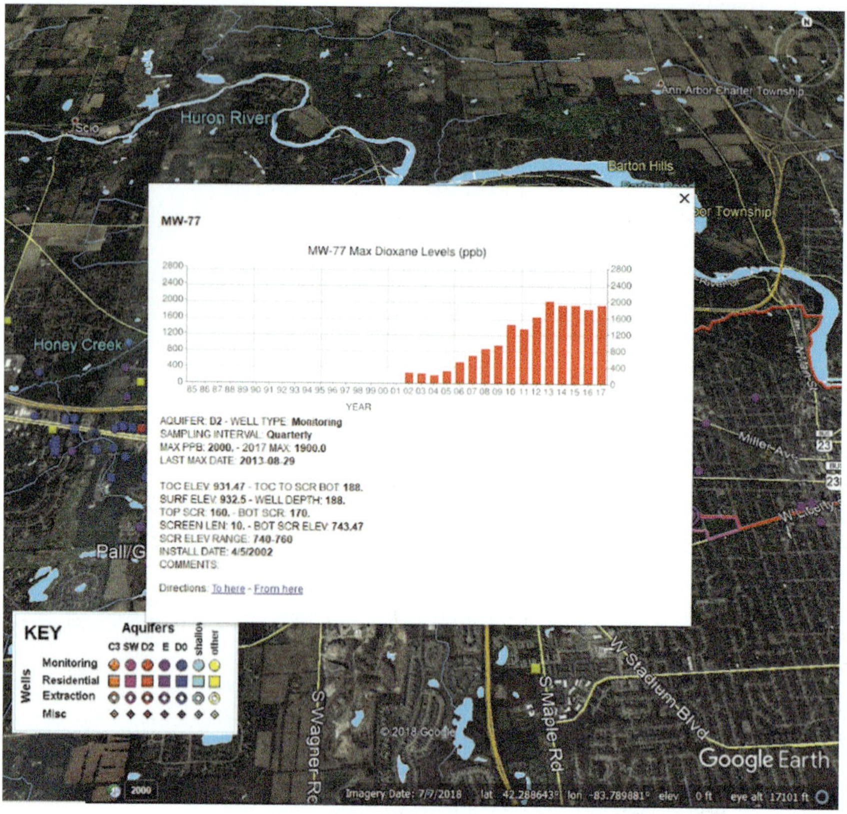

Figure 13.7 *Wells graph. (Source: DEQ–Gelman database (2017). Licensed under a Creative Commons Attribution 3.0 License (http://creativecommons.org/licenses/by/3.0) by Roger Rayle & Scio Residents for Safe Water.)*

A second option shows the maximum dioxane ppb per year as a data table when a well icon is clicked (Figure 13.8).

This second option works offline, unlike the graph option, which requires an online connection to Google's chart function (https://developers.google.com/earth-engine/charts). Both of these well layers come in another layer set sorted by 20 ft. well screen elevations, providing the viewer with the option to see at what elevations dioxane sample points exist or are lacking.

Other layers have been added to provide more in a comprehensive mashup of the site, including

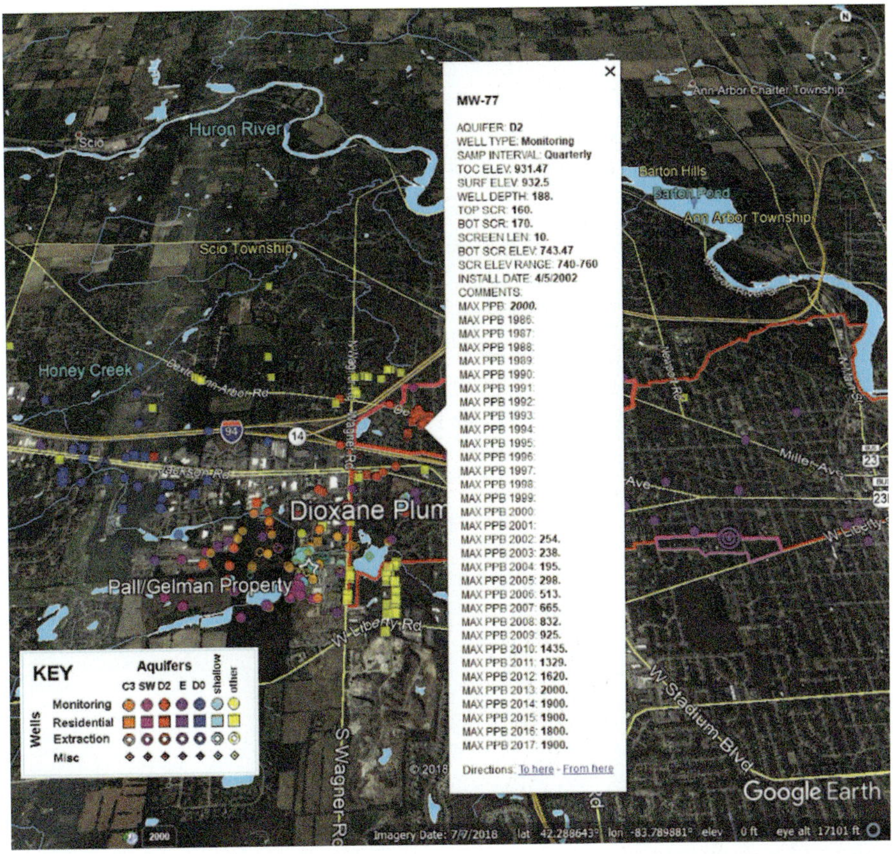

Figure 13.8 *Wells table. (Source: DEQ–Gelman database (2017). Licensed under a Creative Commons Attribution 3.0 License (http://creativecommons.org/licenses/by/3.0) by Roger Rayle & Scio Residents for Safe Water.)*

- Image overlays of contamination plume maps by year (Figure 13.9, Figure 13.10)
- Shape files from local governments showing lakes, streams, drains, watersheds, flood plains, and well restriction zones in the affected area
- Shape files of local government jurisdictions in and near the affected area
- Image overlays of groundwater protection areas for municipal supply wells
- Shape files of groundwater recharge areas
- Image overlays of groundwater flow studies

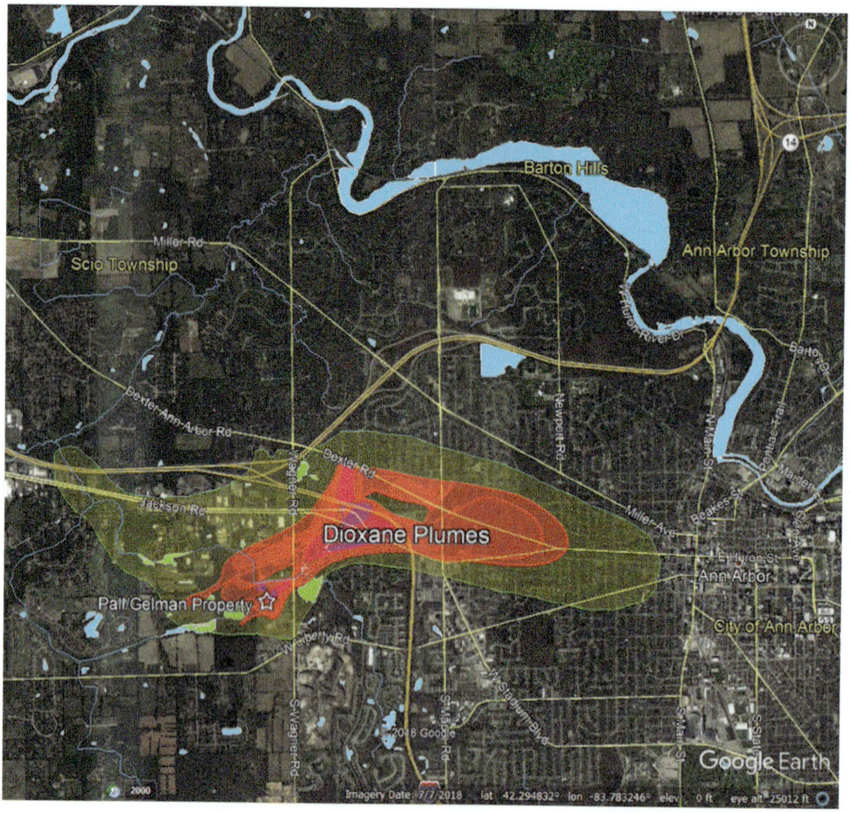

Figure 13.9 *Two-dimensional plume map, 2017 (Gelman plume map, 3rd quarter 2017). (Source: Licensed under a Creative Commons Attribution 3.0 License (http://creativecommons.org/licenses/by/3.0) by Roger Rayle & Scio Residents for Safe Water.)*

- Schematic diagrams of original treatment attempts and subsequent cleanup efforts
- A separate, comprehensive mashup of all available cross-section diagrams showing aquifer lithology in the area based on well borings

With all of these layers available interactively in Google Earth, a viewer can turn layers on and off to generate images to show the available data from whatever viewpoint best depicts the reality of the site and share the result with others. The Scio Residents for Safe Water website contains many links showing current and archived maps and materials on this topic. The author has found Google Earth to be an excellent platform to help stakeholders and others to better understand such complex sites.

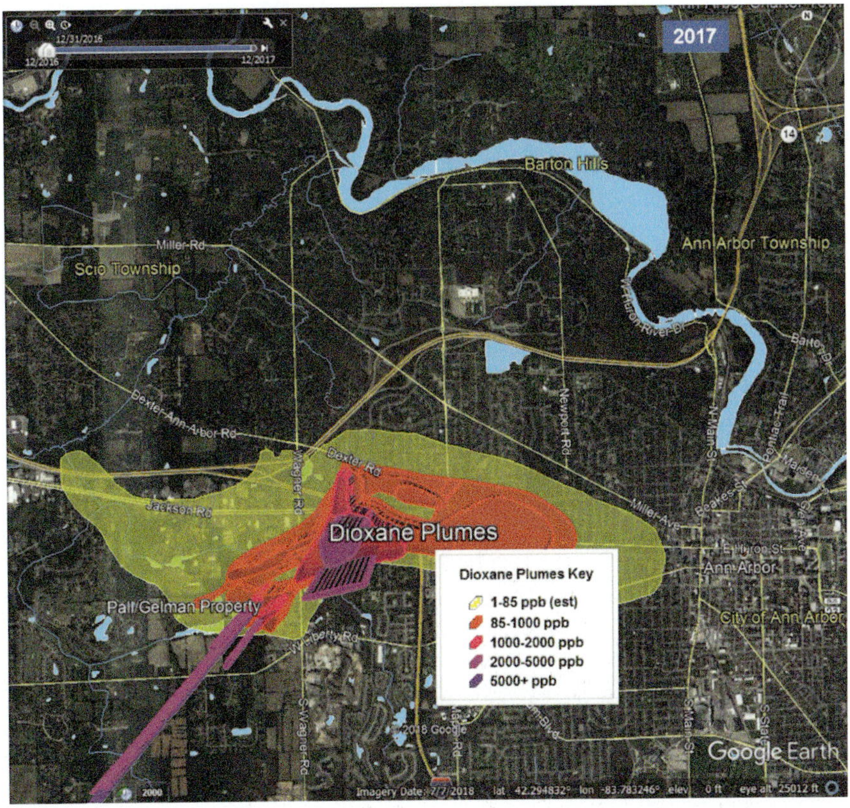

Figure 13.10 *Three-dimensional plume map, 2017 (Gelman plume map, 3rd quarter 2017; extruded by 1 m for each ppb). (Source: Licensed under a Creative Commons Attribution 3.0 License (http://creativecommons.org/licenses/by/3.0) by Roger Rayle & Scio Residents for Safe Water.)*

References

DEQ–Gelman Access Database. 2018–current. Retrieved from https://www.michigan.gov/documents/deq/deq-rrd-GS-DEQ-PLS-DRH-WorkingData_547759_7.zip.

Rayle, R. 2008. Google Earth Applications in a Community Information System: Scio Residents for Safe Water. *Solstice: An Electronic Journal of Geography and Mathematics* 19(1). http://www.imagenet.org.

Rayle, R. 2011. Pall–Gelman plume: A contemporary Google Earth view. *Solstice: An Electronic Journal of Geography and Mathematics* 22(1). http://www.imagenet.org.

Rayle, R. 1986–present. Ongoing field analysis.

14

Pacemaker GEOMAT: My Heart Your Heart

Organization of Efforts Linked by QR Codes

Thomas C. Crawford, Kim A. Eagle, and Sandra L. Arlinghaus

Figure 14.0 *Spatial word cloud summary, based on word frequency, made using Wordle.*

Recycling/Reusing Cardiac Pacemakers and Defibrillators: Rationale

The following is from the My Heart Your Heart project home page.

> There is a great disparity between high- and low-income countries in terms of cardiac pacemaker and defibrillator availability. Each year 1–2 million individuals worldwide die annually due to a lack of access to pacemakers and defibrillators. Meanwhile, almost 90% of individuals with pacemakers would donate their device to others in need if given the chance. The Frankel Cardiovascular Center has been conducting a

> series of research projects to establish pacemaker and defibrillator reuse as a feasible, safe, and ethical means of delivering life-saving therapy to patients with no resources. Throughout this process [My Heart Your Heart] have been engaged with the US Food and Drug Administration (FDA) in order to obtain approval and begin a clinical trial. Our website contains legally vetted forms for making such donations directly or for loved ones to make such donations after death.
>
> We must never forget that at the foundation of each technological breakthrough is the need to improve humanity in all aspects of our society. Undoubtedly, pacemaker reuse can safely and effectively transform a currently wasted resource into an opportunity for a new life for many citizens in our world!

Research projects, and information about research projects, is stored on a permanent basis in Deep Blue, the persistent archive of the University of Michigan Library. The QR code in Figure 14.1 provides a direct link from print to that archive. Please follow our archive about this project to save millions of lives around the world!

GEOMAT Advantage

Deep Blue offers wonderful advantages for scholars at the University of Michigan who have a project that is eligible for inclusion in this digital persistent archive. There are full-time library staff devoted to the maintenance of the digital files and the entire digital archive. This service is a huge advantage in putting the 'moving target' of file formats into an enduring structure in the future, although not all file formats are maintained with the same degree of service. To keep matters simple, storage is in a flat-file structure. There are no subdirectories. One cover page is available to a single collection, and on that

Figure 14.1 *Left: My Heart Your Heart logo; Right: QR code linking to My Heart Your Heart persistent space in Deep Blue. (Source: Reprinted with permission of Director Thomas Crawford (http://www.myheartyourheart.org). Retrieved from https://deepblue.lib.umich.edu/handle/2027.42/109408.)*

cover page one can insert a single image (such as a logo) along with a large amount of text and links to files in the collection. One can upload any number of appropriate images to the storage area (along with text, data, video, and a whole range of digital objects) and create links from the cover page to those stored digital objects.

The special collection for My Heart Your Heart lists a set of published materials, in chronological order, on the cover page for the collection, with links to the associated articles. What the cover page does not show visually, and really cannot, is the acceleration of publication and its clustering over time. It also does not show the geographical distribution of publication and cannot because of the limitation on the number of images on the cover page (which appropriately shows the project logo). The GEOMAT can, however, overcome these limitations. Figure 14.2 displays an associated map, archive, and timeline (GEOMAT) for My Heart Your Heart. The timeline in this GEOMAT differs in appearance from the timeline in the honeybee GEOMAT. In this one, QR codes are inserted, one for each year in the study (Arlinghaus, Haug, and Larimore, 2016). This strategy takes advantage of the persistent nature of Deep Blue. Each QR code is directed to a location in Deep Blue that contains all the currently available materials about My Heart Your Heart within Deep Blue and other locations on the internet. Scan the QR code with a smartphone camera or with a free application to read these 2D bar codes and get taken directly to the appropriate area within Deep Blue.

2009	5 items	Abelardo NS, Baman TS, Eagle KA, Goldman EB, Grezlik R, Kirkpatrick JN, Lange DC, Oral H, Romero A, Romero J, Samson G, Sison EO, Tangco RV.
2010	2 items	Abelardo NS, Baman TS, Caplan AL, Eagle KA, Fuller K, Gakenheimer L, Goldman EB, Kemp SR, Kota K, Kirkpatrick JN, Lange DC, Machado CE, Morgenstern K, Nosowsky R, Oral H, Papini C, Romero A, Romero J, Samson G, Sison EO, Sovitch P, Tangco RV, Verdino RJ.
2011	7 items	Aragam KG, Arlinghaus, SL, **Baman TS**, Brown AC, **Crawford TC**, Duarte C, **Eagle KA**, Feldman D, Gakenheimer L, Ghanbari H, Goldman EB, Hasan R, Kirkpatrick JN, Lange DC, Machado C, Meier P, Menesses D, Oral H, Rivas D, Romero J, Sovitch P, Zakhem NC.
2012	2 items	Arlinghaus, FH Jr., **Baman TS**, **Crawford TC**, **Eagle KA**, Gakenheimer L, Goldman EB, Goold SD, Kirkpatrick JN, Meier P, Oral H, Samson G, Sovitch N, Sovitch P, VanArtsdalen J, Wasserman B.

(Continued)

2013	1 item	Barrera F, Brugada J, Colín L, Gómez J, Iturralde P, Márquez MF, Morales JL, Nava S.
2014	5 items	Arlinghaus SL, **Baman TS**, **Crawford TC**, Desai N, **Eagle KA**, Gakenheimer L, Hagan L, Hughey AB, Kirkpatrick JN, Montgomery D, Oral H, Romero J, Smith CA.
2015	0 items	
2016	0 items	
2017	3 items	Allmendinger C, Alyeshmerni D, Arlinghaus SL, **Baman TS**, Brown A, Carrigan T, **Crawford TC**, Davis S, **Eagle KA**, Emanuel EJ, Goldman E, Hughey A, Klugman N, Kune D, Lautenbach D, Lavan B, Newman C, Oral H, Persad GC, Samson G, Snell J, Sovitch N, Sovitch P, Tandon K, Wasserman B, Weatherwax K.
2018	3 items	Arlinghaus SL, Crane J, Crawford TC, Deering TF, Eagle KA, Haynes D, Marine JE, Pavri BB, Purvis EM, Sinha SK, Sundaram S, Vlay SC, Ward M.

For example, Figure 14.3 shows the result of scanning the QR code for the year 2009. Notice the list of files on the right-hand side. In this image, there are two files one can open. The top one is a simple cover page. The next one is a list of items with links.

Figure 14.4 shows the results of clicking on the second file on the right. That file contains links to PowerPoint displays as well as to articles published in traditional medical journals.

Each of the years since 2009 has its own separate QR code linked to separated areas within Deep Blue. Thus, more files can be added to these areas as time goes on; the QR code remains persistent and will link to that area. So, the GEOMAT is made dynamic by introducing 'spatial thinking' in the form of a QR code—a 'spatial' bar code. Files are added to the persistent areas within

Figure 14.2 *GEOMAT for My Heart Your Heart. The QR codes link to lists of publications, by year. Lists of author names, by year, are arranged in alphabetical order (for convenience in locating an individual). A single name may appear on more than one publication in a given year. To see the author order on the publication, go to the publications of that year. The map shows the names of project individuals associated, for one reason or another, with the Frankel Cardiovascular Center at the University of Michigan. Names of project founders and the director are in bold. The associated QR code, below the map, is a link to the Google Earth .kmz file from which the image was derived. Navigate to the Frankel Cardiovascular Center layer. Turn on the default 3D buildings layer. (Source: IMaGe, ©2014. http://hdl.handle.net/2027.42/108255 and http://hdl.handle.net/2027.42/111740. Used with permission. Link to Google Earth. kmz file, https://deepblue.lib.umich.edu/bitstream/handle/2027.42/111740/MHYH.kmz?sequence=12&isAllowed=y.)*

Deep Blue rather than to the static capture of the GEOMAT presented on pages in a printed book. Thus, at this time, there are no entries associated with 2015 and 2016; but if some become available, they can be uploaded to the appropriate areas of Deep Blue and will appear when the QR code is scanned. The QR codes are important because they add an extra dimension of permanence: they continue to point to materials for a given year, even when those

Figure 14.3 *Page in Deep Blue that appears when scanning the QR code associated with 2009 in Figure 14.2.*

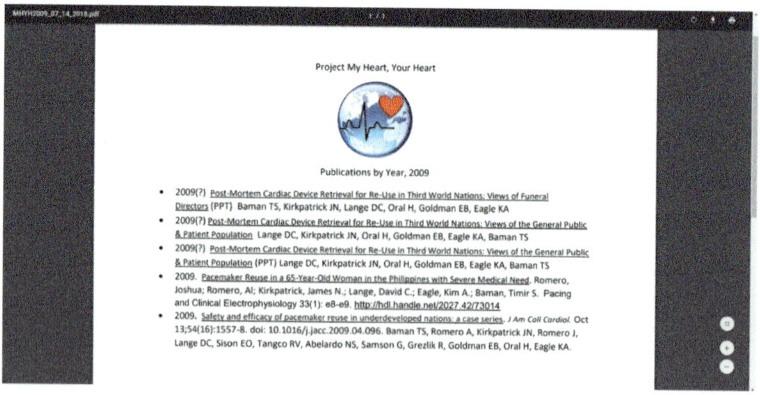

Figure 14.4 *List of publications for the year 2009.*

materials have been updated. Medical professionals, or other interested people, might sit in an airport with this book and a smartphone and read material in the medical environment associated with pacemaker recycling during the period 2009–2018, organized in a GEOMAT format.

Figure 14.3 and Figure 14.4 suggest the style of material that will appear from using the QR code timeline in Figure 14.2. Also, in Figure 14.2, the second column notes the year associated with the corresponding QR code, all presented in a uniform, evenly spaced, calendrical timeline. The third and fourth columns present information about the stored materials so that one can see, at a glance, the pattern of publication and clustering/gaps in the materials. The number of publications per year is quite variable. Is the variation perhaps due to the timing of reviewing and the mechanics of publishing major journals? Or is it due to a lag between doing the field work and writing up the results for publication? Perhaps it, and the gaps, occur for other reasons. A letter in the 2012 archive suggests field trips to Washington, DC to try to alter FDA policy on the time regarding the transport of used medical devices across state lines; clearly this must have been an ongoing issue. Or were there regional outbreaks of disease, such as Ebola in West Africa, impeding certain styles of field progress with consequent slow-downs elsewhere? One important reason to display materials in a GEOMAT format is to cause the reader to ask questions such as these; for it is such questioning that guides research and creates a full picture of the broader environment before, during, and after the study.

The map in Figure 14.2, from which the timeline springs, was made in Google Earth. Again, it evokes questions that can help to guide research. In that map, a number of the authors are shown as having their primary location at the Frankel Cardiovascular Center at the University of Michigan. Where are the other authors? Elsewhere on that campus? In other parts of North America? In universities or elsewhere? In countries outside of North America? What is the level of interdisciplinary connection? What are the merits and difficulties in having colleagues in other countries? Are some selected in association with countries with disease outbreaks for which pacemaker access is of particular importance (e.g., Chagas disease)? Again, having a great deal of information compressed spatially in a single figure offers a significant challenge for research investigation—as suggested here in medical environments.

References

Arlinghaus, S. L., T. C. Crawford, and K. A, Eagle. 2014a. The diffusion of a medical innovation: Visualization using Google Earth. *Solstice: An Electronic Journal of Geography and Mathematics* 25(1).

Arlinghaus, S. L., T. C. Crawford, and K. A. Eagle. 2014b. The diffusion of a medical innovation, part 2. *Solstice: An Electronic Journal of Geography and Mathematics* 25(2).

Arlinghaus, S. L., R. Haug, and A. E. Larimore. 2016. Maps, archives, and timelines: Revisiting the GEOMAT project. *Solstice: An Electronic Journal of Geography and Mathematics* 27(2).

Deep Blue. 2018. Online persistent archive of the University of Michigan. https://deepblue.lib.umich.edu (last accessed August 29, 2018).

My Heart Your Heart. 2018. Online archive. http://www.myheartyourheart.org (last accessed August 4, 2018).

Project My Heart Your Heart. 2018. Persistent digital archive, retrieved from Deep Blue. https://deepblue.lib.umich.edu/handle/2027.42/109408 (last accessed August 4, 2018).

15

Detroit GEOMAT: Chene Street History Study

Scale Transformation, Layering, and Nesting

Marian Krzyzowski, Karen Majewska, Sandra L. Arlinghaus, and Ann Evans Larimore, with input from Anna Topolska, Hannah Litow, Shera Avi-Yonah, Robert Haug

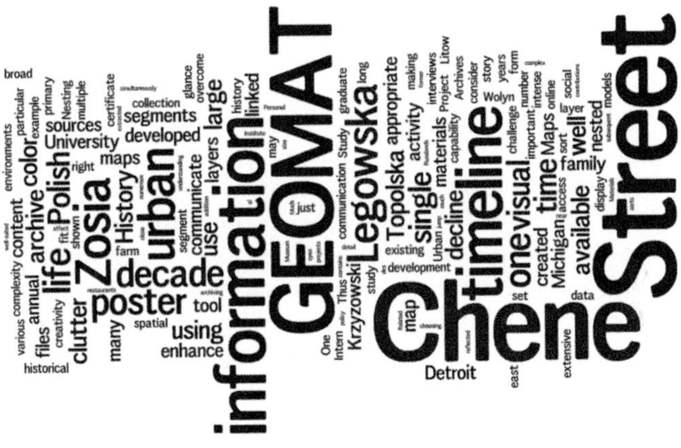

Figure 15.0 *Spatial word cloud summary, based on word frequency, made using Wordle.*

Introduction

To communicate information effectively, choosing the right tool at the right scale is important. We consider an example from the Chene Street History Study, on the east side of Detroit, to illustrate this point, in relation to a GEOMAT, within the urban social environmental context of Detroit (Chene Street History Project, 2014; Chene Street History Study, 2016).

The east side of Detroit, just to the east of the downtown, has a long history of intense urban activity. Migrants from other countries and from the south of the United States, along with religious refugees, have brought a wealth of compact urban development: night clubs, cigar factories, auto plants, 'mom and pop' stores of all sorts, ethnic restaurants reflecting migration patterns, large churches, bars, and various neighborhood activity centers.

Chene Street, a north–south street, served as the central axis for such action. As Hannah Litow et al. noted (2014), in the abstract of a University of Michigan award-winning poster she created with information from a variety of sources (especially from Krzyzowski and Majewska along with some from Avi-Yonah):

> The primary purpose of the Chene Street History Project is to document the 100-year history of this once vibrant and diverse commercial and residential district in Detroit. In the 1920s, Chene Street was the second most commercially-active street in America; now the Chene Street zip code is the least densely-populated area in Detroit. Throughout the U.S., there are many neighborhoods that fall victim to neglect. Residents tend to relocate, taking their families and businesses along with them, in turn leaving the area abandoned. The Poletown and Black Bottom neighborhoods, centered around Chene St., once very highly populated, prominent areas on the east side of Detroit, Michigan are prime examples of this.
>
> This project [Chene Street History Project] aims not only to track the causes of this transformation but also to increase public awareness of the neighborhood's rich history, as a means of generating (and continuing) interest in the revitalization of this community. A community with no history inspires little or no interest. Developing and preserving histories can restore interest and initiate redevelopment. Long term, we hope this project will not only have a concrete impact on people's perceptions of neighborhoods in Detroit but will also provide city officials with a context that they can use to frame their revitalization.

Indeed, outside Fifth Avenue in New York City, Chene (pronounced 'Shane') Street was, for many years, the most intensely developed short urban stretch in the nation (Krzyzowski, 2014–2016; Majewska, 2014–2015). Figure 15.1 offers an abstraction of historical Chene Street complexity. Beyond that rich past and into the 21st century, Chene Street became a wasteland. What happened? As Krzyzowski (director of the Chene Street History Project) has noted, the study of Chene Street, from economic, to social, to spatial impacts, is the 'poster child' for the study of rapid urban decline.

There are many complex factors that twine themselves around these broad concepts, in terms of political actions and so forth, and these no doubt worked together to foster decline. Krzyzowski has spent much of his career documenting the decline by archiving primary sources in numerous formats and

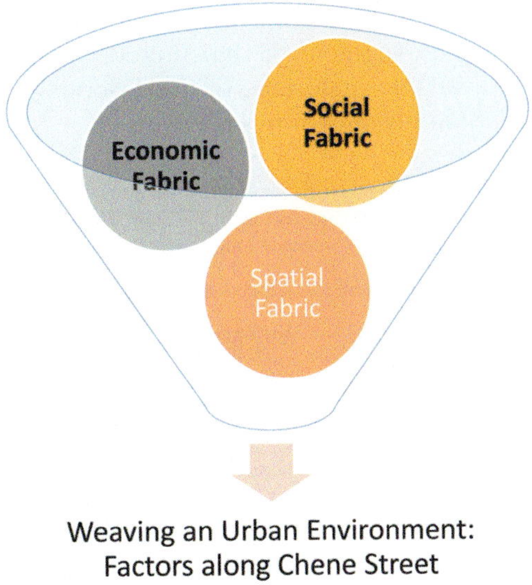

Figure 15.1 *Abstract view of complexity along Chene Street or other regions of intense urban activity.*

working to understand how various forces interacted or failed to interact on behalf of Chene Street and its nearby communities.

A Chene Street GEOMAT

Here, we take a GEOMAT look at one element of the "neighborhood's rich history" that Litow noted: the life of a single individual, Zosia Legowska, who was a fixture for decades along Chene Street. Much of the content of the Legowska story was developed in 2014 by Anna Topolska, who at the time was a graduate history student at the University of Michigan and was doing an internship in association with the Chene Street History Project, for a graduate certificate in Museum Studies at the university. The Legowska story was her primary focus for the certificate. Topolska is bilingual (as are Krzyzowski and Majewska) and was able to access all the Polish sources, including Krzyzowski's interviews with Zosia Legowska, as well as secondary sources published in Polish: she made extensive and creative use of rare materials in an existing archive. Her contemporary vision complemented rigor in understanding urban environments on Chene Street.

The full GEOMAT of the life of Zosia Legowska (with digital files developed primarily by Arlinghaus from Krzyzowski's and Topolska's direct input, along with direct commentary from Larimore and Majewska as well as indirect association with Robert Haug) may be found linked to the QR code below. A clipped snapshot of one segment of that linked GEOMAT, just to give an idea of the styles of available materials, of the fascinating life of this immigrant from Ukraine/Poland is shown in Figure 15.2.

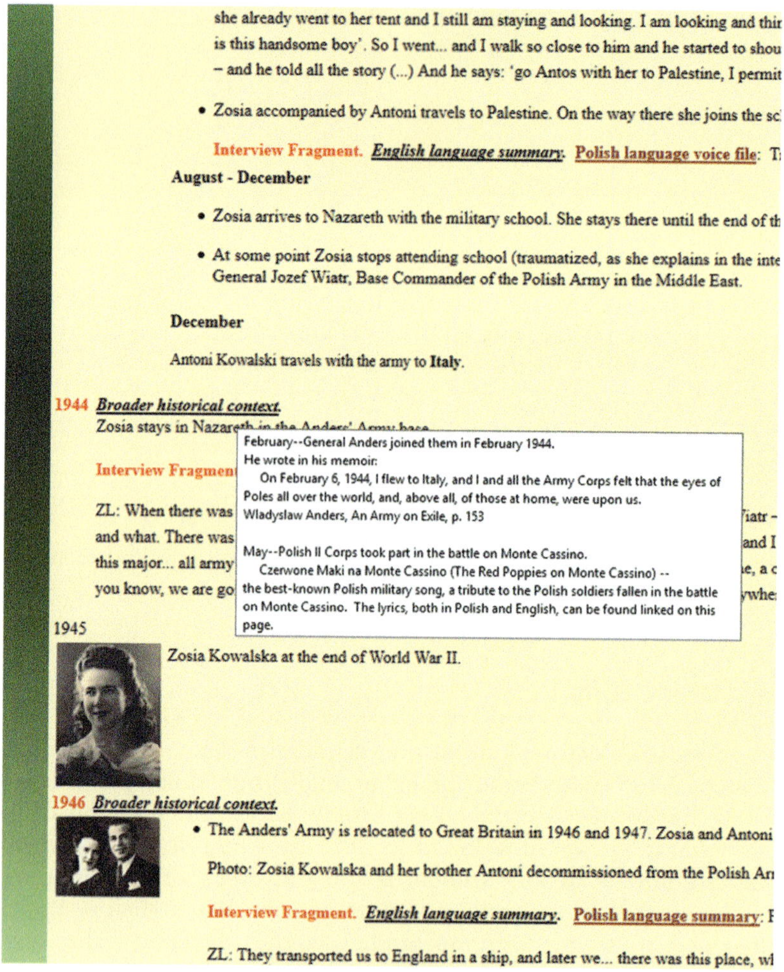
she already went to her tent and I still am staying and looking. I am looking and thir
is this handsome boy'. So I went... and I walk so close to him and he started to shou
– and he told all the story (...) And he says: 'go Antos with her to Palestine, I permit

- Zosia accompanied by Antoni travels to Palestine. On the way there she joins the sc

Interview Fragment. *English language summary*. Polish language voice file: T

August - December

- Zosia arrives to Nazareth with the military school. She stays there until the end of th
- At some point Zosia stops attending school (traumatized, as she explains in the inte General Jozef Wiatr, Base Commander of the Polish Army in the Middle East.

December

Antoni Kowalski travels with the army to **Italy**.

1944 *Broader historical context.*
Zosia stays in Nazar

Interview Fragmen

ZL: When there was ... 'iatr –
and what. There was ... and I
this major... all army ... e, a c
you know, we are go ... ywhe

February--General Anders joined them in February 1944.
He wrote in his memoir:
On February 6, 1944, I flew to Italy, and I and all the Army Corps felt that the eyes of Poles all over the world, and, above all, of those at home, were upon us.
Wladyslaw Anders, An Army on Exile, p. 153

May--Polish II Corps took part in the battle on Monte Cassino.
Czerwone Maki na Monte Cassino (The Red Poppies on Monte Cassino) -- the best-known Polish military song, a tribute to the Polish soldiers fallen in the battle on Monte Cassino. The lyrics, both in Polish and English, can be found linked on this page.

1945
Zosia Kowalska at the end of World War II.

1946 *Broader historical context.*

- The Anders' Army is relocated to Great Britain in 1946 and 1947. Zosia and Antoni

Photo: Zosia Kowalska and her brother Antoni decommissioned from the Polish An

Interview Fragment. *English language summary*. Polish language summary: F

ZL: They transported us to England in a ship, and later we... there was this place, wl

Figure 15.2 *Heavily cropped segment of an existing GEOMAT of Zosia Legowska. http://www.mylovedone.com/Chene/ZosiaPage.html.*

The main thing to note with this GEOMAT is its richness: there are old photos, and there are segments of interviews, conducted by Krzyzowski with Legowska in Polish, both in the original Polish (as .mp3 files) and in English translation by Topolska (mouse over the appropriate places to reveal overlay segments, as in the white box in Figure 15.2). This GEOMAT serves as a fine tool for the serious scholar who wishes to look at life along Chene Street, over more than 50 years, through the eyes of one local resident. For a quick glance, however, information may not jump out at the reader: there is so much information that the superficial appearance of the document has become visually cluttered.

Overcoming Clutter

One way to overcome visual clutter is to layer the information so that individuals are given smaller segments of information in a single glance. This sort of technique is common in making maps, in both conventional hand-drawn maps as well as in electronic geographic information system (GIS) maps. It is a basic principle with visual clutter and it is one that works well in conjunction with the GEOMAT.

Thus, we extracted information, with a broad brushstroke, about the life of Zosia and presented it as a GEOMAT in the format of a rather abbreviated timeline: first as a timeline by decade (Figure 15.3) and then 'nested' as a timeline for one particular decade, the 1940s (Figure 15.4). Each timeline is rooted in an appropriate map. Nesting is one form of layering that can overcome clutter and enhance communication.

The decade GEOMAT was created simply using the 'SmartART' capability of Microsoft Word (on a computer running the Windows 10 operating system). The varying color pattern for the time markers along the timeline was chosen so that subsequent nested layers could be matched with the appropriate color. Thus, the subordinate annual timeline for the period 1940–1949 is color coded with the same color on all annual time markers as the single time marker on the decade GEOMAT (maize). Attention to subtle detail such as this can enhance the ease of communicating using a visual display.

Creative use of the GEOMAT method enhances the ability to communicate. There is no magic formula for making one size fit all: consider the nature of the set of information being mapped and think about ways to best fit the tool to enhance communication: from a single layer created in html (Figure 15.2) to multiple layers, nested or not, using software (of any sort) available to the user (Figures 15.3 and 15.4).

Indeed, other projects deriving information from the extensive archive, in addition to that of Topolska (and others), which were developed about the same time as the linked GEOMAT of the life of Zosia Legowska, include Litow's poster about the rise and demise of a particular jewelry store along

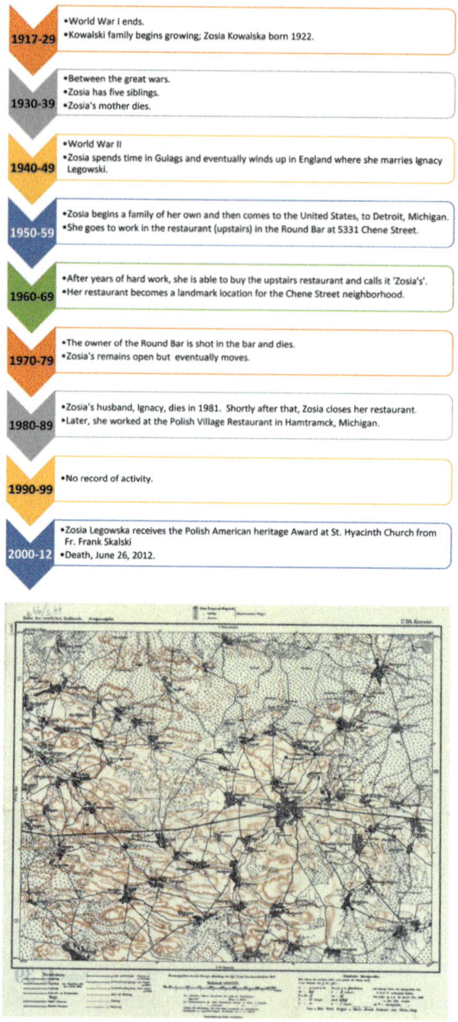

Figure 15.3 *GEOMAT decade timeline. The map shows the birthplace of Zosia, Kozak village, near Korzec town, Wolyn, Wolyn District (Wojewodztwo wolynskie), pre-war Polish Republic (today Ukraine). (Source: Karte des westlichten Russlands, 1917. Retrieved from Mapster (http://igrek.amzp.pl/details.php?id=11216).)*

Chene Street (Litow et al., 2015) as well as Arlinghaus's (2015) more global look at 3D models (many created using archival photos of Chene Street) in the broader Chene Street area as one way to use current technology in support of primary source historical records. It remains for others to rise to the challenge of understanding the past in order to plan for the future. The Chene Street archive is one source of such information; the GEOMAT is a tool to attempt to rise to that challenge. The challenge is important: done well, it can affect

1940
- Soviets began mass deportations of peoples from Polish territories to forced labor camps (Gulags) mostly in Siberia.
- The Kowalski family was taken by Soviets from their home and sent to a variety of Gulags.

1941
- Zosia became separated from her family and experienced considerable difficulty getting back together with them.
- The family was released from imprisonment in order to join the newly-forming Polish army under General Anders.

1942
- The family completes joining the Anders' army where Zosia takes a job in the kitchen.
- Much of the year is devoted to arduous travel, some with the army and some with a military school for girls, eventually winding up in Nazareth, Palestine.

1943
- Zosia remains in Nazareth; sometimes in association with the military girls school, sometimes in association with the army.
- She takes a job at a shop.

1944
- Zosia remained based in Nazareth until the end of World War II.
- She served as a courier for a general and travelled locally.

1945
- End of World War II.

1946
- The Anders' army was relocated to Great Britain. Zosia goes to England.
- In Stratford, Zosia meets her future husband, Ignacy Legowski, an engineer from northern Poland.

1947
- Zosia and Ignacy get married in Birmingham, England.

1948
- No recorded activity.

1949
- No recorded activity.

Figure 15.4 *GEOMAT annual timeline, to nest with the decade timeline of Figure 15.3. Zosia is born in 1922, and in 1923 the family moves to the large Kowalski family farm. It is from that farm that the family is taken by the Soviets, in 1940, to begin their arduous travels that came about in association with the situation in World War II, reflected on this map.*

policy and ultimately create human urban environments that are improvements over previous ones.

Indeed, news of such a challenge broke just as this book was going to press. General Motors announced the closing, by the end of 2019, of its Hamtramck ('Poletown') assembly plant that sits at the top of Chene Street and that, 40 years ago, was built using *eminent domain* powers to clear the traditional neighborhood along and around northern Chene Street. At the time, there was massive local outrage; those feelings have once again been rekindled. Thus, the Chene Street history archive at the University of Michigan, always an important source of regional human/environment/economy information, gains substantial new relevance—as residents rise to meet, and learn about, new challenges surrounding an old issue.

References

Arlinghaus, S. 2015. Chene Street history study, full area. 3D models, 3D Warehouse. https://3dwarehouse.sketchup.com/collection/e24da0f7e786e4bf5a926ed551ef3422/Chene-Street-History-Study-Full-area.

Chene Street History Study. 2016. https://sites.lsa.umich.edu/detroitchenestreet.

Chene Street History Project. 2014. Zosia's page. http://www.mylovedone.com/Chene/ZosiaPage.html. Last accessed 8/4/2018.

Krzyzowski, M. 2014–2016. Personal communication. Field study, photographs, etc.

Litow, H., M. Krzyzowski, K. Majewski, S. Arlinghaus, A. Larimore, and S. Avi-Yonah. 2015. Award-winning poster, Chene Street History Project, University of Michigan, LSA/UROP.

Majewska, K. 2014–2015. Personal communication.

Topolska, A. 2014. Personal communication.

16

GEOMAT Guide: Summary

Study the Past, Understand the Present, Prepare for the Future

Ann Evans Larimore, Sandra L. Arlinghaus, and Robert J. Haug

Figure 16.0 *Spatial word cloud summary, based on word frequency, made using Wordle.*

This guide extracts principles from the examples in previous chapters and also from the imagination as to what a GEOMAT might be and how one should, or might, be organized. Thus, readers see how they can frame the analysis and internet presentation of many different kinds of human events and processes. In Chapter 12 we saw how the evolution of a global environmental pest, the Varroa mite, could be captured in GEOMAT format, which emphasized the paucity of research over a long period followed by an active period of research. What happened in that gap to limit the research process? In Chapter 13, Roger Rayle displayed his vast knowledge of a unique local contamination problem, organized in GEOMAT form, as a critical way to inform the public process and to help mold public policy in a specific locale and consequently offer a guide to others, elsewhere. In Chapter 14, an archive of an important emerging medical technology, cardiac pacemaker reuse/recycling, explored the novelty of inserting QR codes in the GEOMAT timeline to keep current the archive of associated human events and medical advances. And, in Chapter 15, the GEOMAT served to capture a specific actor from the Chene Street History Study and organize her world travels, activities, and related human events (from materials presented in both English and Polish languages), as one key to understanding change in a particular urban area.

Using the unique properties of the internet, the GEOMAT can articulate vast amounts of archival documents and images in time and space and make them accessible easily and quickly. The application of a bit of imagination to examples such as these permitted us to create sets of principles helpful in guiding GEOMAT formulation in a multitude of contexts.

A GEOMAT

- Can be used for many kinds of investigations, whether for political, detective, intelligence, and other investigative work or for academic research questions.
- Integrates analysis and synthesis. We can then use the internet not just for information/data (i.e., facts) but also for creating knowledge and understanding.
- Can aid conflict resolution by creating web architectures for all sides of a conflict, articulating large amounts of documents and other data in a single website.

GEOMAT: Generalized Conceptual Structure

Two general goals for the GEOMAT web architecture are as follows:

- To make available as much data as possible
- To do that with as much accuracy and precision as the archived data/documents allow

A parallel, related concept is to do so in a manner that provides access to the procedure that is straightforward in execution in developing, as well as developed, nations.

Four ecological statements form the GEOMAT framework that can be used at varying scales from the global to the local.

- Every event occurs at a single time in a single place.

- Some events are landmark events that change our planet's history and geography irrevocably. The Yukon gold rush, the opening of the Suez Canal, and the eruption of Mt. Saint Helens are examples.
- In each event both human actors and environmental actors are involved.
- Their actions are recorded and preserved in archives, landscapes, and settlements.

When creating a GEOMAT case study, both human data and environmental data can be integrated into an ecological whole, erasing the divide between 'human' and 'nature'. The GEOMAT design enables ecological thinking in webs of connections through time-space in branching patterns.

The Core Matrix: Timeline and Location Map

The core of the GEOMAT web architecture is a matrix juxtaposing two web pages locating the central landmark event in time and space: one a vertical timeline showing the dates of the event being analyzed, the second an index map of places/sites significant to the event. In these two core components are set link anchors of text or other symbols that access data and documents articulated to key times and places. The insistence on the vertical nature of the timeline reflects the way the English language is written (i.e., horizontally). When the timeline is vertical there is no overlapping of labels associated with adjacent timeline entries (obvious adjustments of timeline orientation might be made for languages written in other orientations).

A GEOMAT case study articulating large amounts of data and documents necessarily reflects the web architecture builder's own interpretation as presented in the framing and arrangement of the original materials. Readers, however, can engage the original documents directly as well and through that process come to their own conclusions, interpretations, and readings. This is the beauty of providing original data and documents immediately available electronically.

The great value of the GEOMAT web architecture for research lies in its capacity to include vast numbers of archival materials. This is a technical advantage over print media that quickly become too cumbersome for a researcher to interact with; indeed, the presentation of a GEOMAT in print format, as is done here, underscores the importance of viewing it, additionally, in electronic format.

Because of the articulation of particular archival documents in electronic proximity, an investigator may be able to discern relationships between facts and patterns produced by arrangements of facts much more easily than with print. The conversion of crime statistics, for example, from lists arranged by time of reporting to animated maps will show spatial patterns not otherwise evident. Health statistics arranged alphabetically by place when mapped may reveal patterns of occurrence and transmission otherwise invisible.

The GEOMAT Content's Structure

For the full power of a GEOMAT case study to be realized, two sets of intersecting components need to be included. First is a set of broad categories of substantive data that form an ecological whole: these eight systems interact simultaneously at any place on the earth's surface. Human individuals in groups are embedded in these systems and act through them. Identifying the essential actors, human or environmental, that have produced a particular event enables us to analyze how these systems interact in time and space.

Categories of Substantive Data

The set of broad categories that form an ecological whole is as follows:

1. Climatic and weather systems, including the water cycle, regional climatic seasons, and extreme weather conditions
2. Terrain and topographic formations, including mountain chains and plateaus, their geomorphology, and the watersheds of rivers and streams
3. Distribution of flora and fauna and other natural resources, such as mineral deposits, including changes like the movements of plants and animals both domesticated and wild
4. Population settlements, villages, towns, and cities, and movements such as urbanization and migration, including their routes of transportation and communication
5. Family formation, operation, and reproduction through generations
6. Political institutions' formation, operation, reproduction, spread and decline
7. Social institutions' formation, operation, reproduction, spread and decline
8. Economic institutions formation, operation, reproduction, spread and decline, including those of land use systems.

These eight systems interact simultaneously at any place on the earth's surface. Human individuals and groups are embedded in these systems and act through them. Analyzing how these systems interact to produce a particular event enables us to identify the essential actors, human or environmental, that have produced the event.

Format Categories

Second is a set of *format categories*, as follows:

1. Maps showing various features of areas where landmark events took place
2. Calendrical timelines showing the sequence of different kinds of events at appropriate timescales, pop-out timelines, and zoom-in timelines (nested timelines)
3. Identification of specific events, especially landmark events that have irrevocably changed the situation being chronicled in a case study
4. Biographies, accounts, and images of human actors—corporate group agents contributing to the sequence of events and their institutions and organizations
5. Reports about and images of other actors, such as plants, animals, landscape features, terrain, mineral deposits, productive land use, weather, and climate
6. Arrays of archived primary documents—governmental and private, records of messages exchanged, oral history accounts, and historical and contemporary images
7. Accounts of, and documents and images from, contemporary settlements, such as cities, towns, villages, resorts, and rural areas, farmed and wild
8. Documents and images from archeological sites and records of past settlements

These eight kinds of information categories serve to alert the investigator to the need to search out information in different conceptual frameworks and formats.

To conclude, the GEOMAT template, with its core matrix timeline map, makes it possible to ensure accuracy, degrees of precision, comprehensive documentation, ecological reasoning tracing out web connections, and the presentation of conflicting stories of events positioned in relation to the agreed-upon facts. The GEOMAT is an analytical, synthesizing, and presentation methodology that innovates by utilizing the unique properties of the internet—its global reach, its website construction architecture, its storage capacity, and its instantaneous linking capability—to create knowledge and understanding of geographical changes.

Community and Academic GEOMAT Study Group Operational Procedures

GEOMATs may be used by individual researchers for analysis, synthesis, and presentation. They may also be used by teams working on the solutions to conflict situations, on the revealing of the facts about a puzzling or little-known set of events, or on the organization of fragmentary materials in complex situations.

To utilize the GEOMAT web architecture method to contribute to conflict solving and other types of team work, the following work structure may be useful.

- Form a collaborative small study group: scholars and stakeholders with differing perspectives on the conflict to be studied. The chair should be outside the issue at hand but knowledgeable about collaborative procedures and process.
- The study group's procedures and process need to be designed to be fair, equitable, consensual, and open to scrutiny.
- The study group should establish a charge/mission statement and a calendar of operations.
- The study group should seek funding for specialist personnel to develop the GEOMAT/complex of GEOMATs under study group supervision.
- Study group calendar: The work would be organized corresponding to participant calendars. A standardized work calendar should be established: data and document collection, web page design decisions, and text and image composition for individual webpages. An electronic library of data and documents outside the GEOMAT itself would be developed.
- Public participation at regular intervals: the developing GEOMAT should be put on the internet in a read-only format for a few weeks following each meeting. There should be bulletin boards for public comment.
- A contact address would be provided for the submission of data and documents to the study group for the study group's decision to accept or reject. To be accepted, all data would need the time and location of data and the public name and address of the contributor to be verified and confirmed. Metadata (authorship, date, purpose, agents, and conditions of creation) would be required for any documents or data submitted.
- For permanency of the GEOMAT development as it moves along, there would be regular backups stored in a safe location. Copyright and similar items would of course be honored.

Once these operational procedures are coupled with the scholarly concerns involving mapping, timelines, text, and a variety of historical and geographical matters, the GEOMAT evolves into its full and useful form!

Bibliography

Hillier, A., and A. K. Knowles. 2008. *Placing History: How Maps, Spatial Data, and GIS are Changing Historical Scholarship*. Redlands, CA: Esri Press.

Knowles, A. K., ed. 2002. *Past Time, Past Place: GIS for History*. Redlands, CA: Esri Press.

Gregory, I. N., and P. Ell. 2009. *Historical GIS: Technologies, Methodologies, and Scholarship*, Cambridge Studies in Historical Geography. Cambridge University Press.

Based on an original paper by the same authors, available at http://www.geomats.org.

V

Story Maps, 2010s

This section begins with an explanation of different types of story maps before presenting the Varroa study as a story map. Again, we compare and contrast this method for mapping with the ones of the previous three sections in the environmental context of the plight of the Varroa map. Both sample chapters employ cascading story maps as these appear as a direct evolution of the GEOMAT. A final comprehensive chapter takes a careful, hands-on look (written in a personal tone) at various types of story maps to reinforce the intertwined nature of all that has gone before in a practical manner.

17

Types of Story Maps

Joseph J. Kerski

Figure 17.0 *Spatial word cloud summary, based on word frequency, made using Wordle.*

Web Mapping Applications

Web mapping tools typically present the user in a web browser with a series of tools that accomplish certain functions, such as measurement, symbology, and analysis, along with an interactive map canvas, and the ability to save, embed, and share. Web mapping tools allow the map author to decide upon the content, symbology, scale, classification method, projection, and other map elements. The author can usually make a copy of content that someone else has authored, modify it, and add to it, or the author can create completely original content from his or her own spreadsheets, desktop geographic information system (GIS), fieldwork, or other data sets. The content comes from web mapping services that are on servers that are either offsite, such as those in the Amazon, Microsoft, or Google cloud, or onsite, such as the organization's own servers. Web mapping platforms include those from Esri (e.g., ArcGIS Online), Mapbox, QGIS, Leaflet, and others. Web mapping involves client-side and server-side tools and functions. Many web mapping interfaces allow the user who is not signed into the system (an 'anonymous user') to perform certain functions, and a user who is signed into the system (a 'named user') to perform additional functions. For example, ArcGIS Online allows the anonymous user to modify the map by adding data, symbolizing and classifying mapped features, measure distances and areas, find locations, and other functions. ArcGIS Online allows the named user to save the map, share the map, create web mapping applications such as story maps, and perform spatial analysis on that data. The named user is granted certain rights to the organizational account in which they are a part of: Some users are administrators, who can invite users to the organization and grant access to specific tools to them. Other users are publishers, who can publish data layers and perform analysis. Other users can only view data and make temporary changes to maps. ArcGIS Online is a manifestation of the Software-as-a-Service (SaaS) model discussed earlier: It is a GIS running in the cloud, with data, tools, and services available to the user.

Web mapping *applications* are created from this type of interactive web GIS platform. Each platform has its own interface, which usually includes a 2D or 3D map canvas, a list of map layers, and a set of tools, all running in a web browser. A web map is a map created using this interface. If that map is shared with others, others see the same interface. They can also interact with the map, and add data to it, symbolize or classify it differently, and save it into their own content area.

By contrast, a web mapping application is given as a product, a presentation, or embedded in a final multimedia experience. Web mapping applications can therefore be considered the 'final product' that the map creator delivers to the client, whether the client is a group of students, professors, patrons, customers, a specific audience, advisors, board of directors, or others. The recipients of the web mapping application typically do not change the application, though many are modifiable through code customization or the addition of content through crowdsourcing. The recipients thus interact with the application, read the story, change the scale, pan to different areas, and read and watch the text, videos, and photographs—by and large, they *consume* the content.

Story Maps

Story maps are specific types of web mapping applications. They combine interactive maps, multimedia content, and user experiences to tell stories about the world, from local to global scale. The stories can be about a current event, a subject (e.g., natural hazards or population changes), an issue (e.g., plastic pollution in a local river), or anything else that takes place on, under, or above the Earth's surface (though story maps do exist about the Moon, planets, stars, and other celestial bodies). Story maps can be created on several different platforms and technologies. These include story maps JavaScript (Storymaps.js), Mapstory (Mapstory.org), and custom-made story maps by specific organizations (e.g., those from the *New York Times*). This chapter will focus on Esri story maps. The reasons for such a focus is because Esri story maps are based on the ArcGIS Online cloud-based infrastructure, which gives the maps a much greater variety and volume of data, tools, and sharing capabilities than the other platforms. Esri story maps are thus connected and an integral part of a much larger GIS platform. Esri story maps are hosted by Esri as cloud-based entities. However, the creator of the story map can also opt to host them on his or her own website. Story maps are open source and can be downloaded and customized. Story maps are optimized for any device, including laptops, tablets, and smartphones.

Types of Story Maps

Story Map Tours

Story maps include an array of 'app types' that provide for different ways of interaction and thus different experiences. These app types include map tours, map journals, cascade, shortlist, series, swipe, spyglass, crowdsourced, and basic. The easiest story map app to create, and as of 2019 still the most common type of story map, is the map tour. As shown in Figure 17.1, a map tour includes a single map, which could have many layers, along with a carousel of drawings and images, along with a larger image showing whatever the user is examining in more detail. The user can navigate through the story via the map or via the images. The map shown also touches on a theme of this chapter that story maps are enabling organizations to communicate their mission and vision that may not have used maps in the past. Here, the City of San

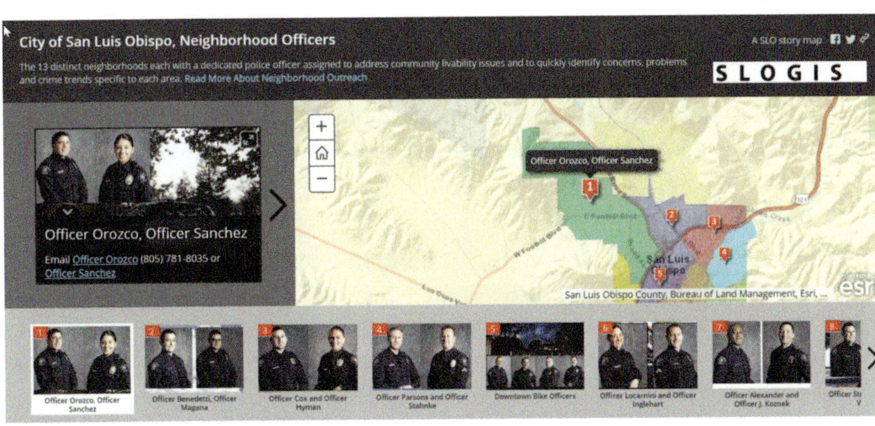

Figure 17.1 *Example of a map tour type of story map. (Source: Esri and City of San Luis Obispo. © 2018 Esri. Used with permission.)*

Luis Obispo Police Department uses story maps to introduce the community to the officers protecting that community, putting real faces and names to the department, and reinforcing their role as partnering with the community to reduce crime and protect its citizens.

Another example of the use of a story map tour is a book entitled *International Perspectives on Teaching and Learning with GIS in Secondary Schools* (http://denverro.maps.arcgis.com/apps/MapTour/?appid=5f86647b1e8e491aadaece6345927f2a). The authors used a map tour to feature the 33 chapters in the book, each focused on a different country, as a way of increasing interest in the ways GIS is used in education (Kerski et al., 2013).

Another story map showcases the work of the Esri Young Scholars (http://denverro.maps.arcgis.com/apps/MapTour/index.html?appid=a383612f79354488929beabcd266cd77), and because the scholars come from all over the world, the map tour was a suitable communications technique to display where they are from and their work. The map also shows enhancements that can be easily made to story maps. Here, under each photograph of the young scholars is a caption, and the caption is hyperlinked to a poster of each scholar's research results.

Story Map Journals

Story map journals allow for a greater variety of multimedia to be incorporated than do map tours. Journals also allow for a greater amount of text narrative. A map tour allows the user to access just one online map. It may contain many map layers, but the layers are still bound into one map. A map journal, by contrast, allows the user and the map creator to access an unlimited number of online maps. The user moves through the story through a series of sections, one at a time, or skips to specific sections, or uses the section arrows to navigate (as seen on the right of Figure 17.2). At any point, just as with any story map, the user can also zoom, pan, make specific map layers visible, and perform other functions on the web maps in the story map. Another example of a map journal story map highlights the citizen science work done as part of the BioBlitz effort in Hawai'i Volcanoes National Park (http://story.maps.arcgis.com/apps/MapJournal/?appid=45867f2ae46e4587afb8e7c7b343b9b8).

Another example of the effective use of a looping video, 2D and 3D maps, and a current issue is of the Elwha River in Washington State, in this story map called "A River Reborn" (http://storymaps.esri.com/stories/2015/river-reborn) about the removal of two century-old dams, and the environmental, social, and legal story leading up to the removal and after the removal.

A map series allows content to be accessed through a series of tabs. The tabs can contain different map layers about a common theme, as is the case in Figure 17.3, where one tab shows geology, another shows forest type, and still another shows land and ecology.

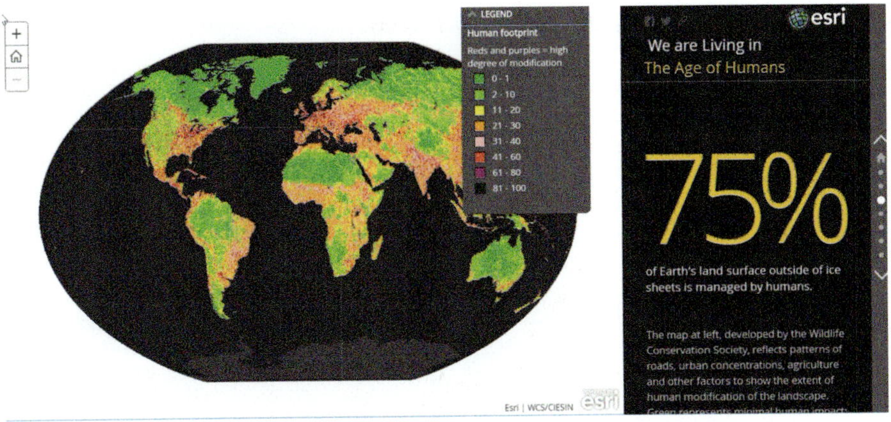

Figure 17.2 *A map journal as a part of the series "Living in the Age of Humans." (Source: Esri. © 2018 Esri. Used with permission.)*

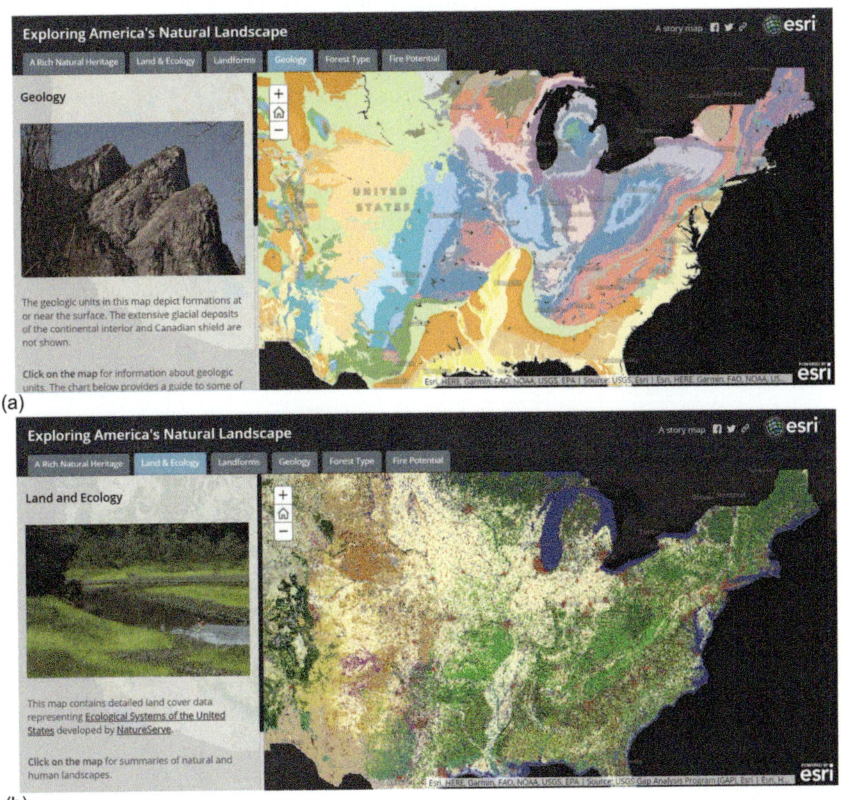

Figure 17.3 *Map series about America's natural landscape. (Source: Esri. © 2018 Esri. Used with permission.)*

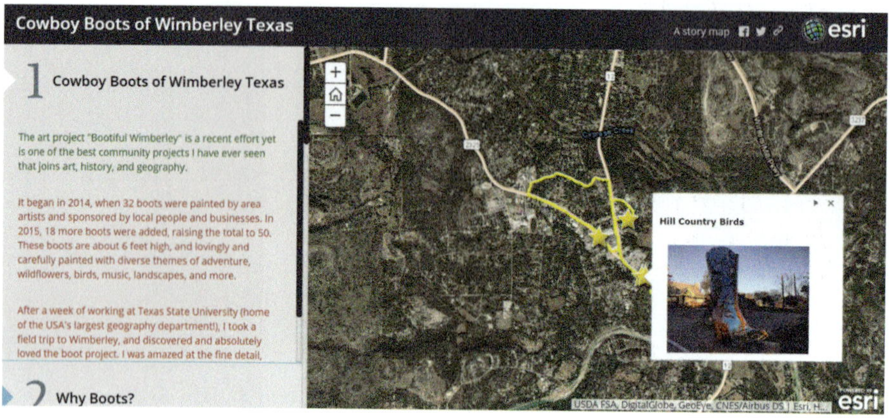

Figure 17.4 *Map series about a public art project in Wimberley, Texas. (Source: Esri. © 2018 Esri. Used with permission.)*

A map (Figure 17.4) created by the author of this chapter features video and photographs of a cowboy boot public art project in Wimberley, Texas (http://denverro.maps.arcgis.com/apps/MapSeries/index.html?appid=76f8dafc79674ebdba40f3dc9ac6d6c8). Each boot stands about 9 ft. tall and tells its own story about an aspect of Texas—ecoregions of Texas, birds, history, and more. This story map contains a web map of historical sites in Texas, videos of selected boots, photographs of boots, and a virtual tour that the map reader can use offsite or take into the field to plan his or her own walking route.

Story Map Swipe and Spyglass

The story map swipe and spyglass app types invite the user to explore different content through a squeegee vertical bar effect or through a magnifying glass effect. The content may show different themes, such as land use versus zoning in a community, or different vintages for the same area, such as satellite imagery or topographic maps, or changes in demographic characteristics over time. Figure 17.5 shows a modern satellite image of San Francisco with the spyglass allowing the user to "see back in time" to a map of the city from 100 years ago, to assess where and to what extent the city has changed. These changes include the construction of buildings and freeways, and the fill and subsequent development in the Marina District in the northeast quadrant of the city.

Shortlist Story Maps

Shortlist story maps are so named because a series of thumbnail images are displayed in a list gallery to the side of the map. As the map is zoomed and panned, the images change to reflect the map extent that the user has set.

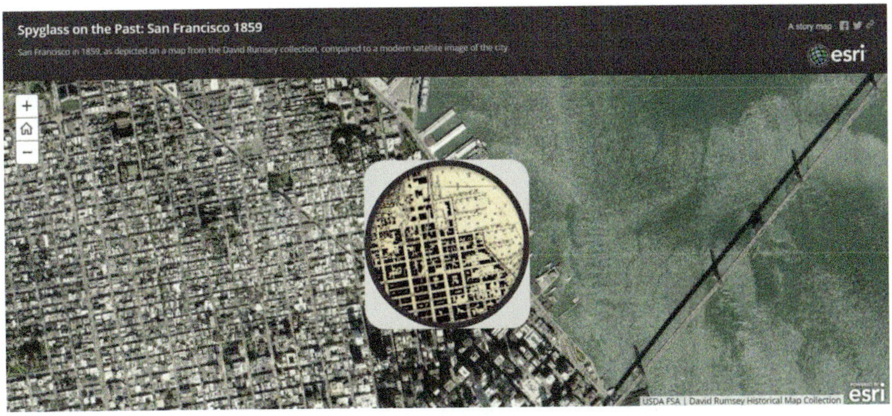

Figure 17.5 *Spyglass story map of San Francisco showing a modern satellite image and historical map of San Francisco.*

Figure 17.6 shows a shortlist map of colleges in California by the California higher education system. It also reflects the theme of maps telling a story; in this map, users have instant access to further information about each institution of higher education but also where that institution is located, helping them to perhaps make a decision on where to apply.

Cascade Story Maps

Story map cascade apps provide an interactive, flowing experience for the user. The example shown in Figure 17.7 that focuses on Syrian refugees illustrates another aspect of story maps—that they allow for 3D content as well as 2D maps to be incorporated. This map also begins with a looping video

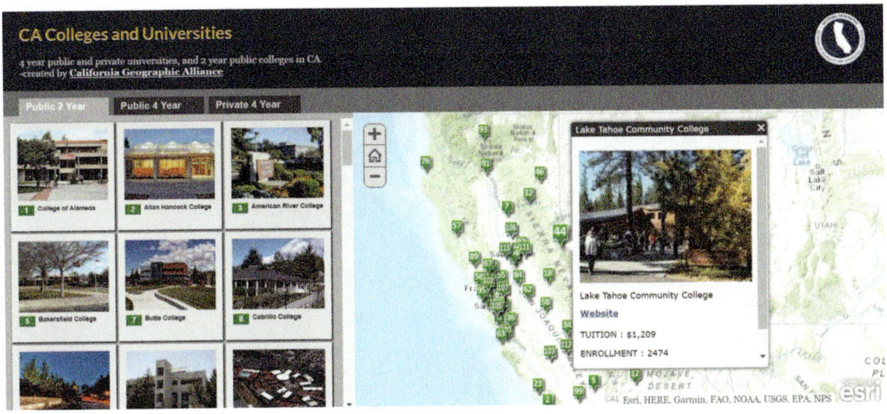

Figure 17.6 *Shortlist story map showing institutions of higher education in California. (Source: Esri. © 2018 Esri. Used with permission.)*

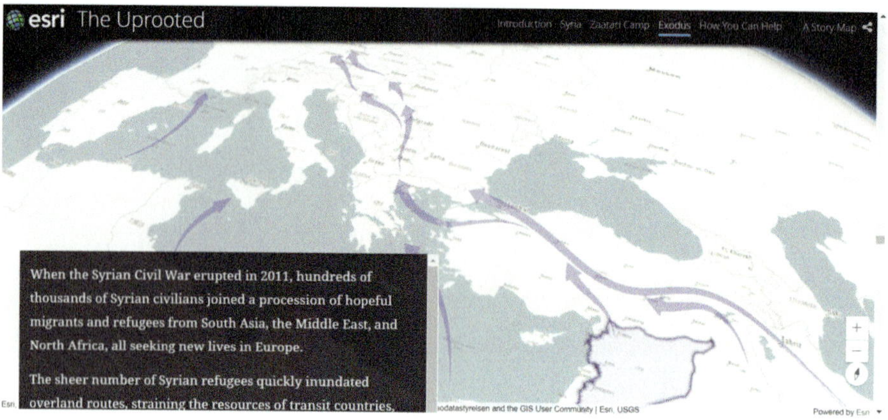

Figure 17.7 *Part of the cascade story map explaining the story of Syrian refugees, with 2D and 3D maps, images, videos, and text.*

designed to quickly engage the map reader, as well as some cartography in the form of arrows that shows customizations possible with story maps.

A crowdsourced story map asks for users to input a short caption and a photograph on a theme asked for by the story map creator. Figure 17.8 shows a map created by the author of this chapter for an online course, to foster discussion about landforms, but also to help the course participants to get to know each other through the text and photographs they post. Participants are provided with a URL for the map, and input their information via a button, here using "Add your landscape." This type of story map is open for input until the author of the story map turns off the editing capabilities. The author also has the option to review each point input before approving it. While the

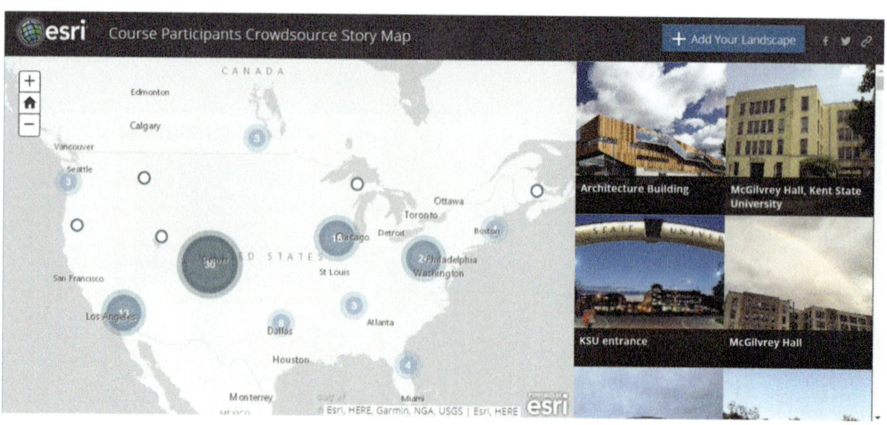

Figure 17.8 *Crowdsourced story map asking for input about the landscape in the participants' locations. (Source: Esri. © 2018 Esri. Used with permission.)*

crowdsourcing map was removed from the story map apps zone in late 2018, it is still available for use. Crowdsourcing story maps can also be created using the citizen science tools Collector for ArcGIS and Survey123.

Story Maps Gallery

An ideal way to begin accessing story maps is to visit the gallery (https://storymaps.arcgis.com/en/gallery/#s=0). The gallery can be browsed by the type of story map app, by subject (business and economy, nature and conservation, science and technology, and others), by industry (education, transportation, and others), by format (standard, customized, embedded, linked, and collections), and by author (Esri story maps team, Esri employee, Esri partners, Esri international, and the non-Esri user community). The content of the gallery is periodically rotated to show innovative, new, and important maps and topics. The gallery also only contains a small fraction of the total number of story maps. By the end of 2018, over a million story maps had been published; the gallery only contains a few thousand maps. Because story maps are based on the ArcGIS Online platform, another way to find other story maps is to search ArcGIS Online (www.arcgis.com). Another way is to search blogs, videos, Twitter, LinkedIn, the Esri GeoNet community (https://geonet.esri.com), and other web resources.

Map Presentations and Story Maps

Stories can be told in a variety of ways with Web GIS tools. Let us compare two examples of communicating the same theme using these tools and maps. On the theme "Why use GIS in education?", the author of this chapter created an ArcGIS Online presentation (http://www.arcgis.com/apps/presentation/index.html?webmap=6e06d858c1ea4888859c03494c9df6ad) (Figure 17.9).

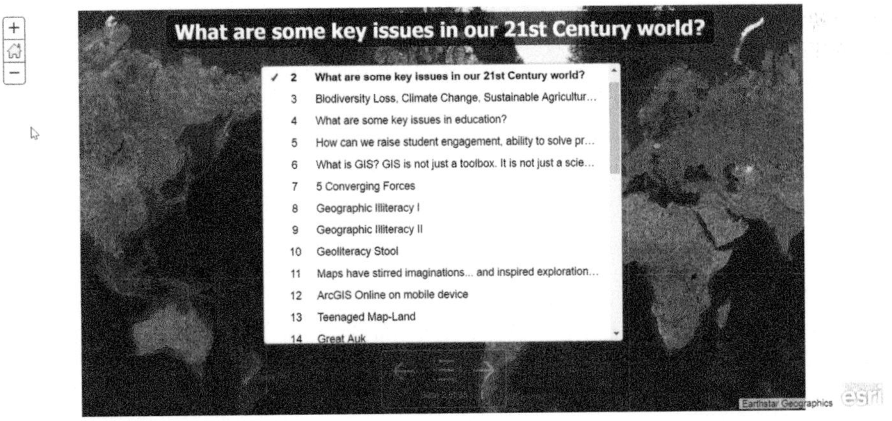

Figure 17.9 *A presentation on GIS in education and society, created and shown as a map presentation. (Source: Joseph J. Kerski.)*

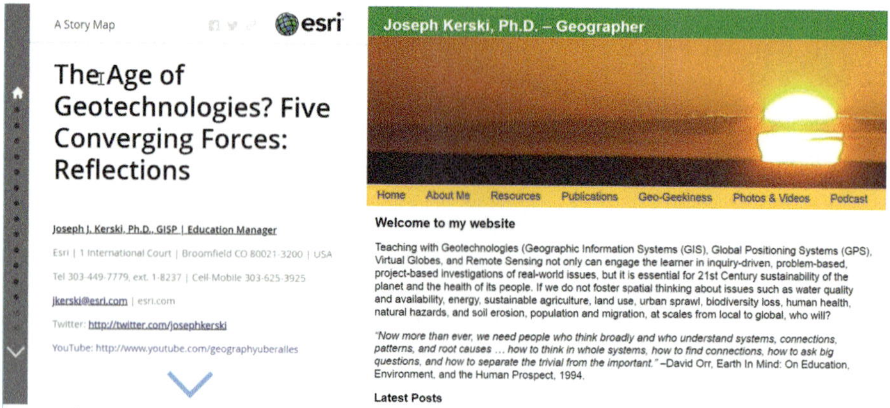

Figure 17.10 *The same theme as Figure 17.9 (GIS in education and society) but shown here as a map journal story map.*

This same theme is also presented as a story map (http://denverro.maps.arcgis.com/apps/MapJournal/?appid=b55fb50a2cef41b8908baa4c376088b8) (Figure 17.10).

Each of these two ways of communicating has its advantages: Both are interactive and are stored in ArcGIS Online. If the author shares them with the public, as these are, they can be accessed at any time, anywhere. They are also both optimized to be shown on any device from laptop to smartphone. The map presentation requires less time to set up but is less powerful: Only one presentation can be associated with each web map, and each 'frame' can contain only text and an interactive map. A story map, while requiring more time to create, can point to multiple web maps and include videos and other multimedia.

Advantages of Story Maps in Education

Advantages of using story maps in education include that they offer an engaging, rich way to teach content. They also foster technical skills in GIS, multimedia, and working with a wide variety and formats of data, such as GIS files, map services, images, and video. They also foster critical and spatial thinking. They nurture skills in data organization, and in ethics and permissions. They provide an excellent way of grappling with types of questions such as, "Can I use that image in my story map?" and "Do I have the right to do so?" They provide a convenient means for instructors to assess students' or colleagues' work, simply via a URL that is sent to the instructor. They also support research and collaboration, since they can be shared and edited. They also foster oral communications skills as students use them to present the results of their work, and when they do so, they encourage discussion by their peers, because the presenting student can interact with the maps as their peers ask questions. In other words, the story maps act as fully interactive presentation tools, rather than static slides of content. Perhaps most importantly, story maps enhance understanding of the phenomena or issue at hand, through the power of maps, charts, text, and multimedia.

Story maps are opening the door in primary, secondary, and university education to departments that did not previously use any geotechnology. One example is work that the New York University conflict and peace studies program is doing with their study of languages spoken and the people behind those languages on a particular subway line from Queens to Manhattan. This initial foray into GIS, through story maps, creates an opportunity for the students and faculty in that department, and others like it, to see more of what they can do with GIS. This is different from the traditional means that a university department begins to use GIS, but again, Web GIS is changing many paradigms. At the primary and secondary level, educators teach fundamental concepts in history, biology, geography, civics, and language arts with story maps. Students use and create story maps using Chromebooks, iPads, and Mac and PC-based laptops. They use story maps to present the results of their research. Web GIS has helped raise the number of schools in the United States using GIS from 700 to 7,000 in just a few years (Kerski, 2018), and the number is growing worldwide as well.

Creating Story Maps

Three methods exist for creating story maps: First, story maps can be created using 'builder' tools. These builder tools emulate a 'wizard' type of workflow that is common in other software, involving a sequence where the user is presented with options, selects 'Next', chooses options on the next screen, selects 'Next', and so on until the final product is created. These builder functions enable a map author to create one without any background in coding or in GIS (though these skills certainly help with making more powerful story maps).

Second, story maps can be created from new or existing web maps in ArcGIS Online. The method involves creating the map in ArcGIS Online, then sharing it, and then creating a story map web application.

Third, story maps can be created by downloading the configurable apps, customizing them, and hosting the results on one's own website. Downloading the configurable apps involves over 100 files, most of which are JavaScript, which is the language of ArcGIS Online, but the user typically adjusts only a small handful of files pointing to his or her own content and publishes the set of maps to a website of their own choosing.

Maps versus Apps

While story maps are part of ArcGIS Online, it is important to discuss the difference between web maps and web mapping applications. Both are useful for instruction and research, but they often have different purposes and

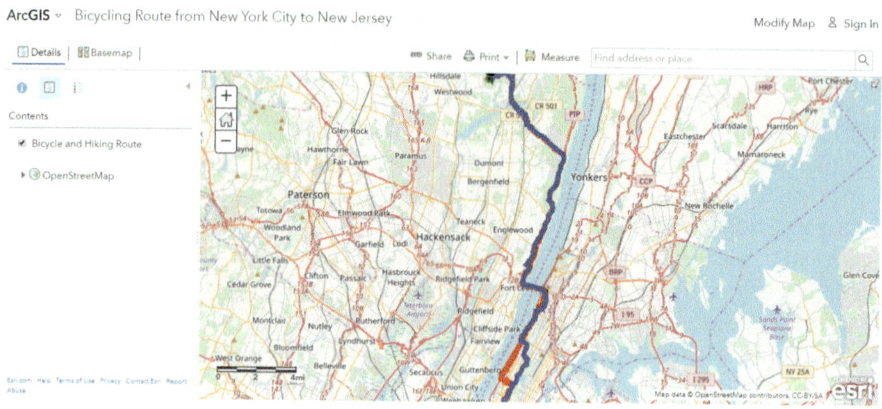

Figure 17.11 *Web map in ArcGIS Online showing the author's bicycling route from New York City to New Jersey.*

are created differently. A web 'map' is stored in the ArcGIS Online cloud and contains the full interface and set of tools with which map users can interact. Figure 17.11 shows a web map for a bicycling route that the author took from New York City to New Jersey following a geography conference. Note that the full interface for ArcGIS Online is shown, including the base map options, measure tool, modify map, contents, and legend. This map could be saved by any ArcGIS Online user, which places the map into his or her own account, symbolized differently, with additional content added as desired. Note that if this is done, the original author's map is not changed, but rather a copy of it.

A web mapping application or 'app' is stored in the ArcGIS Online cloud, just as a map is, but an app contains a specialized or reduced set of tools for the map user to interact with. Figure 17.12 shows a web mapping application for the same bicycling route the author took from New York City to New Jersey,

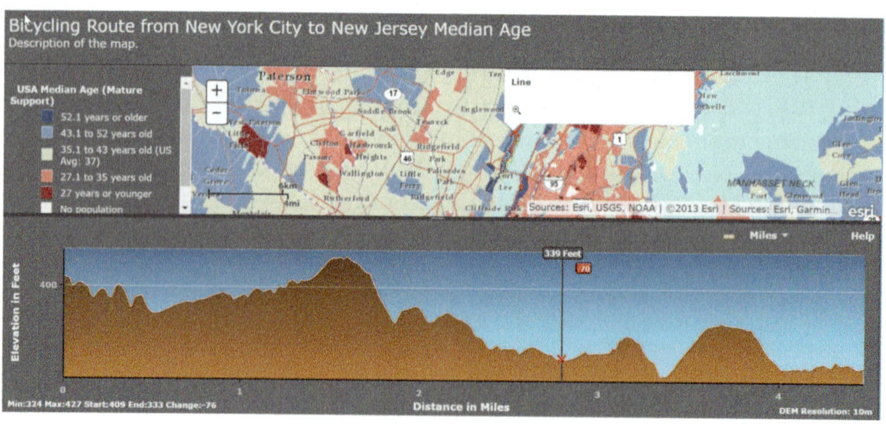

Figure 17.12 *Story map showing the same bicycling route as a story map, with the median age of the neighborhoods and a 3D profile of the route.*

with a 3D profile that the user can select to any segment along the route, and with median age for each neighborhood the author bicycled through.

Therefore, story maps are *applications*. (It just sounds better to say 'story maps' instead of 'story apps' but story maps are actually apps.) A story map is a type of web mapping application that is stored in the ArcGIS Online cloud, and has a specific set of tabs, text, photographs, video, and audio capabilities presented to the map user. Think of story maps as the final product that a map reader will see, whereas a web map contains the components for the final product. Thought of in another way, web maps are the Lego bricks, and the story map is the final Lego structure (glued and varnished; that is, it cannot be altered except by the original author).

Figure 17.13 shows a different web story map application for the same bicycling route the author took from New York City to New Jersey, this time with tabs showing different types of census demographic and tapestry behavior data for each neighborhood.

Consider the Audience

Maps are powerful means of communication. Because story maps are meant to be shared with a wider audience than would perhaps read a journal article or even a book, consideration of audience is even more important with these maps than with other avenues of expression. First, *consider* your audience: are they GIS professionals, people in your industry or field, other professionals, students, or the general public? Your story map's layout, content, and symbology all should be created with your end user in mind. The map user's background and the experience that you want that map user to have should

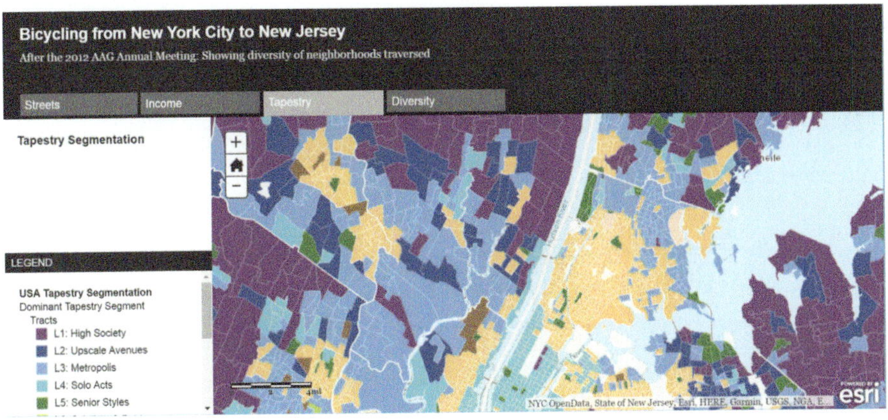

Figure 17.13 *Series story map of the area of the bicycling route with tapestry neighborhood demographic and behaviors.*

dictate the type of story map you create—such as a map tour versus a cascade map. At the time of this writing, you cannot change the type of story map you create midway through the process. Once you commit, for example, to a map journal, you have to continue making a map journal; you cannot change to a series map halfway through the process. Sometime during 2019, however, the story map app types and tools will be presented to the user in more of a free-flowing, creative environment, where the user can choose from a variety of app types and tools at any time. But even when that happens, consideration of the audience will still be critically important.

Second, once you determine which is your primary audience for your story map, make sure that you build the story map to *connect* with that audience. Create a compelling title and lead your audience into your map content with a focused story that unfolds as they explore.

Third, no matter what the makeup of the audience will be, make the maps easy to read. Part of this guideline is to strive for simplicity. The maps in your story map will likely be different from the maps you are working with as a GIS analyst, for example. Consider the maps in Figure 17.14. The map in the upper right shows desert, grassland, and rainforest quite clearly and would be perfect for a story map on world biomes, for example. The map in the lower left is a typical map that a GIS professional works with every day on the job. While useful for analysis, the many layers, symbols, and tools would be confusing to the reader of a story map. Avoid complex symbols that may

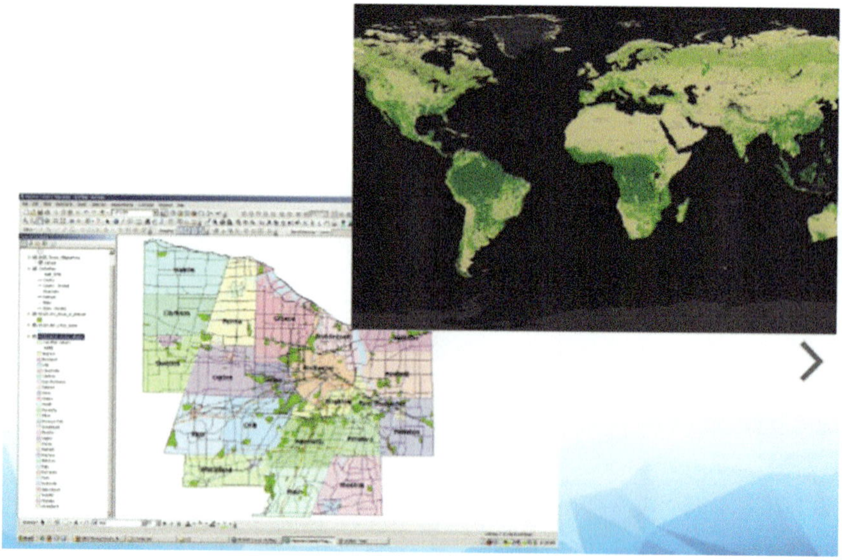

Figure 17.14 *Complex maps (e.g., that in the lower left) are typically used in GIS analysis, but simpler maps (e.g., that in the upper right) may be more suitable for story mapping.*

be difficult to understand and slow to draw in a web browser. Also consider your base map carefully, make certain that it *enhances* the content and does not *interfere* with it. For example, the topographic base map is something that most map creators and map users enjoy viewing and understand, but for many stories, its colors and density of features may occlude the primary story themes and messages.

In addition, lead people in with an appealing and interesting title. Titles such as "On the Brink" invite the map reader to investigate a map (Figure 17.15), and the mosaic of animal images further draws the reader in. The story map "Zoo Babies" is one of the more popular story maps ever created, describing the young animals living in the Washington, DC zoo. Consider how "Zoo Babies" draws the reader in much more than a title such as "Young Animals in the Washington, DC Zoo," because who doesn't love baby animals? The accompanying photo is equally intriguing, and this map features each animal as it lives in the zoo, and also as it lives in the wild, inviting the reader to learn

(a) (b)

Figure 17.15 *Two story maps with titles and images that draw in the map reader.*

about each animal's natural habitat. Consider also the thumbnail and description for the story map carefully, in part because these are often all that the potential audience will see on social media feeds.

Technical Guidelines in Story Mapping

Story maps offer the instructor at many different educational levels—from primary school to secondary, from technical college to university, from lifelong learning to informal settings such as after-school clubs, museums, and libraries—a wealth of teachable moments and an opportunity to foster technical and other skills. These moments and opportunities include those focused around media fluency, geospatial technology, ethics, and broader societal issues.

First, story maps offer an excellent means to foster organizational skills. Using the 'My Stories' zone in ArcGIS Online is a way for users to manage their own content. My Stories allows the story map authors to edit the appearance of their story maps, add content to them, and modify them in other ways. It also allows the map author the ability to make certain all links are working. In other words, it offers capabilities for the curation of the map, which is critically important in the rapidly changing world of the web. My Stories also allows the map creator to edit the map underlying the story maps—adding or changing the map layers, changing the base map, altering the scale or extent, changing the symbology or classification method, and so on. Another guideline is to give context of where the map and study area is located, through an index map or via another means, and the study matters. Often, story maps place the map reader in a large-scale study area, and the map reader has no idea where the study area is or why the issue is important.

Working with Web GIS usually requires many tabs in a web browser to be used. Keep tabs on these tabs: In which tab are you logged in and actively editing? Which tab contains the latest version of your map? Along these lines, do not use the back button in the browser during an active edit session. There is no auto-save, so the back button will erase any edits since the last time the map was saved. Be sure to use folders in ArcGIS Online to organize working with web maps. Any GIS, including Web GIS, includes enough moving parts; keep these parts manageable by being as organized as possible. As indicated earlier, model good practice of the permitted use of imagery on the web. Use your own content, creative commons, or non-copyrighted content (e.g., most of US federal agency data). If you do not have permission, seek it and do not use it unless you have permission to do so.

Finally, be mindful of image size and map content complexity. While bandwidth is improving, 100 images in a map, for example, each of which is 12 Mb in size, will slow down the experience for the end user, as will 10,000 points on the same map.

Finally, follow these three guidelines when working with story maps.

1. The story maps available for use in education are rapidly expanding—the types of apps, the spatial data available, and the tools.
2. The story map platform is rapidly evolving. Therefore, practice being a lifelong learner and do not get "used to using" a certain user interface, because it will likely change. But, each time the platform evolves (which is usually quarterly), it becomes more powerful and easier to use.
3. Good planning makes for a good story map. Just like going out into the field, plan ahead; think about the audience, message, data, and map types that you need for your story. Sketch it out on paper or digitally before creating the map itself.

References

Kerski, J. 2018. Why GIS in education matters. *Geospatial World*. https://www.geospatialworld.net/blogs/why-gis-in-education-matters.

Kerski, J. J.,A. Demirci, and A. J. Milson. 2013. The Global Landscape of GIS in Secondary Education. *Journal of Geography* 112(6): 232–247.

18

Varroa Story Map: All Together Now

Sandra L. Arlinghaus and Diana Sammataro

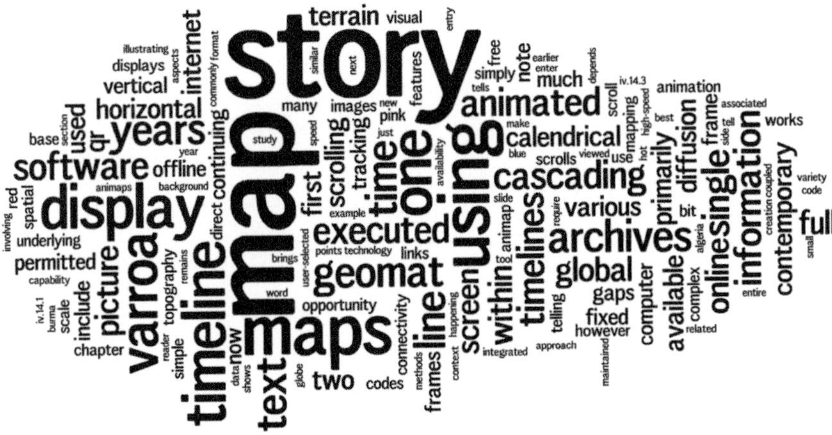

Figure 18.0 *Spatial word cloud summary, based on word frequency, made using Wordle.*

One important element of the GEOMAT involves the platform on which it is built: one that does not require much use of the internet or high-speed connectivity and can easily be built offline and simply uploaded to a server using only a small amount of connect-time. The Esri story map, on the other hand, provides a powerful online tool for also telling a story that links maps, archives, and timelines. It does so with contemporary technology and the power of the internet coupled with high-speed connectivity.

Tracking Varroa

Tracking Varroa is a difficult task. How can one be sure when the earliest sighting of a moving target actually happened? Different sources will have different dates for the 'first' occurrence of Varroa in various regions. However, what they all have in common is a similar spread of pattern. It is capturing that spread that is of interest here (the reader interested in variations in detail might enjoy sources mentioned in Chapters 4, 7, and 12).

One simple, but useful, map combined a number of ideas to use an online tool to track Varroa over time and tell the reader a bit about its global spread (Figure 18.1a). Hot spots on the map show, in coordination with a small user-controlled slide bar as a timeline, when Varroa was first found in various locales. Click on a hot spot and a bit of information pops up. Pull down the arrows under the timeline to learn a bit more about Varroa. Clearly, this map tells part of the story of the spread of Varroa. When the slider is pulled all the way across, the map is covered with hotspots (Figure 18.1b). A useful extra pull-down states the names of regions that are Varroa free. The clutter,

Figure 18.1 *(a) Early and (b) late stages of the spread of Varroa. (Source: Used with permission of the author/creator, William Blomstedt (http://wblomst.com/varroa).)*

however, makes it difficult in a single map frame to see what is where. There is only so much information one can put on a postage stamp.

Unification of Concepts: The Cascading Story Map

To remove the clutter by introducing scrolling across the map was therefore an attractive option. The cascading story map provided that sort of flowing option as it also provided an interactive map experience for the user. Varroa data slides in and out in layers against fixed backdrops. As much information as is desired can be introduced in a manner that reduces clutter. We recast the Varroa GEOMAT as a cascading story map so that the reader has a side-by-side comparison of the two similar approaches and can make a reasoned decision about which approach is best suited to his or her technological and scholarly needs (Figures 18.2a–h). We select only a portion of the available frames to display in this print format. In the digital format from which they were derived, all appear within a single bounded area of the screen and slide in and out.

To view the entire story map, rather than just selected static images, use the QR code in Figure 18.3. Enjoy the dynamic story that it tells. Note the compression of large amounts of information within a single computer screen: maps, archives, and timelines.

When the researcher has a full range of contemporary computing capability coupled with high-speed internet access, the cascading story map is an attractive choice for assembling maps, archives, and timelines. However, when he/she has only limited internet capability, the GEOMAT offers a similar, but not as compressed, opportunity using software that is, and has been for many years, widely available around the globe. Both can be updated in a straightforward fashion, by simply adding new materials rather than having to redo materials from the beginning in order to incorporate new items. The advantage of the story map is that the process of updating can draw on the power of a computing cloud and on easy collaboration around the globe.

Discussion of the Varroa Example

Each of the four major sections of this book—animaps, 3D mapping, GEOMAT, and story maps—has begun by illustrating the methodology of that section using the Varroa example. Which method was 'best' for illustrating the Varroa case study? The answer is "It depends." It depends on the target audience, on the emphasis that is placed on communicating various aspects of the study, and on the priorities for ordering those aspects.

- When animaps were used,
    - The information on diffusion was emphasized.
    - The display embraced the full global distribution, which was visible in a single frame.

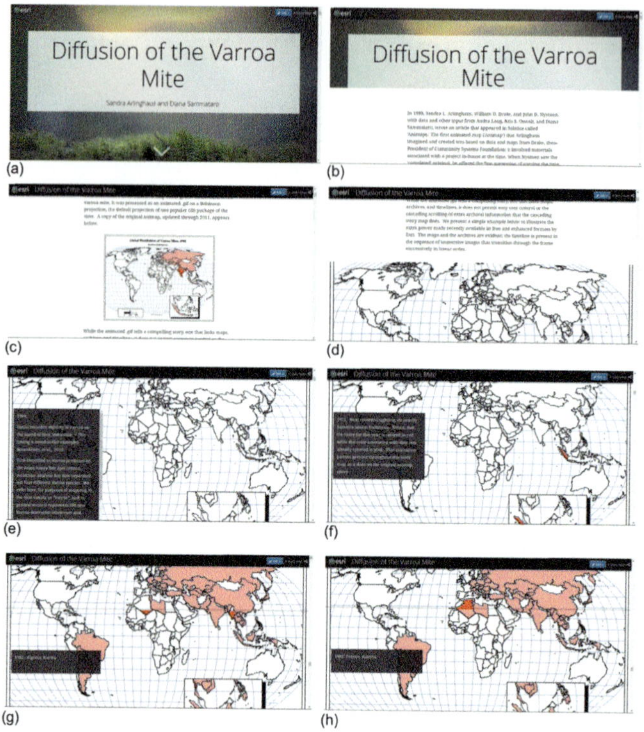

Figure 18.2 *(a) Title screen leading to a cascading story map. Enter by clicking on the down arrow (when viewed on the computer). (b) As one scrolls down, the screen remains fixed (note the text beginning to cover the splash screen), and text or other objects enter from the bottom. (c) Scrolling down further shows continuing text as well as the animated Varroa map (which remains animated within the story map when viewed on a computer). (d) At the end of the text setting the context, the animated map is enlarged and used as a background for the rest of the frames of the story map. (e) Scrolling down further still keeps the background animated map fixed, and the timeline, with associated text, rolls over the map. Here there is material about 1904 coming in on top of the animated base map. (f) Continuing to scroll down brings in subsequent years; note the gap between years. Time gaps are maintained as one scrolls down, so that this desirable aspect of the GEOMAT is maintained while keeping the whole display compact within one computer screen. Only the materials superimposed on the map are scrolling in; the animated map and the entire display stays fixed within a single frame. Scrolling also brings in associated years in the animated base map so that base map and cascading text always correspond. (g) Notice the faint blue horizontal line. As one scrolls down, this line moves. Here, the horizontal line is tracking the entry of Algeria and Burma (as it was called in 1981) into the picture—red below the line, entering the picture; white above the line, not yet in the picture. (h) Continuing to scroll into the next year now shows the horizontal blue line tracking the color change from red to pink: from 1981 to 1982.*

Figure 18.3 *QR code to link to full story map (https://www.arcgis.com/apps/Cascade/index.html?appid=ec089d6d8cdc48cb8cfa33dc22d6b2b9.)*

  - Time gaps were displayed subtly, in the temporal spacing between animation frames.
  - There was no linkage to underlying terrain, topography, historical events, or other related information.
  - Creation of the animap can be executed offline; it does, however, require some software to form the animation, and there may be a fair amount of manual effort.
- When 3D mapping was used,
  - The information on diffusion was presented in relation to underlying terrain and topography.
  - Animation of single 3D frames could be employed to make an animap that now included underlying terrain and topography.
  - Timeline native to the 3D software could be invoked within the display.
  - Level surfaces could be introduced to illustrate terrain effects in a direct, visual manner.
  - The full globe could not be shown in a single frame.
  - Creation of the display was done mostly online.
  - Updating was not fully automated.
- When GEOMAT was used,
  - All of the features from the first two methods could be integrated.
  - A vertical calendrical timeline permitted the clear and direct display of time gaps.
  - A vertical calendrical timeline permitted an opportunity to include the story of what was happening in various years, using text, links, QR codes, images, or other displays of contemporary availability.
  - They were initially created primarily offline using commonly available, or free, software. After offline creation, they were uploaded to online server space, thereby conserving valuable connect-time.
  - Manual updating is usual.
  - There is no capability for platform-related collaboration.

- When cascading story maps were used,
  - All of the features from the first two methods could be integrated.
  - A vertical calendrical timeline permitted the clear and direct display of time gaps.
  - A vertical calendrical timeline permitted an opportunity to include the story of what was happening in various years, using text, links, QR codes, images, or other displays of contemporary availability.
  - They were executed primarily online using contemporary technology.
  - The capabilities of cloud computing permit enhanced automation in terms of updating and collaboration.
- Thus,
  - Animap display works for a simple visual display focused only on the diffusion over time and empty geographical space; it can be simply executed and created primarily offline and embraces full global reach.
  - 3D mapping display works for more complex visual displays linking diffusion to terrain and related features, but not at a full global scale. It is executed online using free software.
  - GEOMAT works for complex story telling involving maps, archives, and timelines and can include all in the first two points at user-selected scale. It is executed primarily offline and uses free or commonly available software.
  - Story maps work for complex story telling involving maps, archives, and timelines and can include all in the first two points at user-selected scale. It is executed primarily online, using a variety of software options. It has an important advantage in terms of collaboration.

19

Tree Canopy Story Map: Health Impact Assessments Inform Community Tree Planting and Climate Adaptation Strategies

Matthew Naud with input from Sandra L. Arlinghaus

Figure 19.0 Spatial word cloud summary, based on word frequency, made using Wordle.

Urban forests are important community assets. Beyond the visual appeal of tree-lined streets, many communities are recognizing that street trees contribute to managing storm water runoff, an especially important value in areas measuring an increase in precipitation due to climate change. Other benefits include providing shade that reduces urban cooling costs and counteracting problems associated with the urban heat island effect. In many urban areas, lower income and vulnerable populations tend to live in areas with lower tree canopy cover (Gill et al., 2007). Therefore, planting additional street trees in urban areas offers a relatively low-cost option to mitigate heat-related public health challenges. This chapter explores the process one community used to explore climate adaptation strategies in lower-income communities. Adaptation strategies in this sector include increasing tree quantity, planting tree species adapted to new climate conditions, and diversifying species type.

Based on 2012 work by University of Michigan graduate students (Bergquist et al., 2012), we began with the hypothesis that poorer neighborhoods are less canopied than wealthier neighborhoods. If this is true, then designing a tree-planting program that focuses on these lower-income, less canopied neighborhoods will provide the same storm water benefits and add the new co-benefits of shading to reduce heat island effects and lower energy costs to neighborhoods most at risk to climate change. We began to discuss whether tree planting could be a low-cost adaptive strategy with the right analysis and opportunity to provide insight to the forestry program. At the same time, the State of Michigan Department of Community Health was looking to support Health Impact Assessments (HIAs) under a cooperative agreement with the Centers for Disease Control and chose to support our urban canopy analysis project. This story map shares our spatial thinking and begins to explore the environmental context for urban equity issues involving urban forestry, tree canopy coverage, and vulnerable populations with public health disparities. Future work can explore the extent of improved canopy coverage, actual reductions in energy bills normalized with weather, and reduced incidence of heat-related diseases with community health data.

HIAs evaluate the potential health impacts of a project or policy and provide recommendations to increase positive health co-benefits and mitigate negative health impacts. Characteristics of HIAs include a broad definition of health; consideration of economic, social, or environmental health determinants; application to a broad set of policy sectors; the involvement of affected stakeholders; explicit concerns about social justice; and a commitment to transparency (Smith et al., 2014). The following work uses existing data from the census to infer health vulnerabilities and integrates those findings with tree canopy measurements to identify neighborhoods likely to have higher health vulnerabilities and lower canopy cover.

Real-world analysis is frequently different from laboratory analysis: there are few controls in the laboratory of the real world. First, we present, in a traditional manner, a process that was used to consider a complex set of data in support of two different themes: tree canopy and public health. From that we assume an association between tree canopy and public health vulnerabilities for different neighborhoods in Ann Arbor, Michigan, by georeferencing the data sets on Esri GIS maps. These maps, and their associated information, are used to form a cascading story map atlas.

In Chapter 17, we saw various styles of story maps on the same topic. In this chapter, we once again use a cascading story map (as in Chapter 18), but this time the scrolling process is used to move from one targeted topic to the next, rather than to move along an embedded timeline. The Varroa case employed the story map to track a single health threat over many years. Here, we use it instead to track multiple disease vulnerabilities as a temporal baseline for future study. These data accumulated on inferred disease vulnerabilities were accumulated over time. They were gathered and analyzed as part of an interdisciplinary study, multi-dimensional across various municipal and academic arenas, funded by a grant (Smith et al., 2014).

The story map offers an opportunity to collapse a set of maps, spread out over a number of report pages, into a compact single-frame arrangement in which the analysis scrolls across the maps, as appropriate. The atlas is presented as a foundational document for an archive of information associated with inferred health vulnerabilities and tree canopy cover by census tract (neighborhood) for the City of Ann Arbor.

The cascading format was selected for its interactive flowing map capability. The 'spyglass' style of a story map might also be effective with the given data set; however, it might well be better used as more data are accumulated, since the density of data is initially quite sparse. A swipe story map was also an attractive alternative; however, the boundary files of the original maps did not overlay properly, so that forming any form of animation, be it a simple animated map or a swipe map, would have produced a poor visual result and inaccurately placed map features.

Two Themes: Tree Canopy and Public Health

Most homeowners and renters know that large mature trees help to shield their homes from excess heat in the hot seasons of the year. The interior of the residence is more pleasant; fresh air from the outside is picked up by breezes and moved inside. Mainly, though, it is the shade from the trees that improves quality of life in our homes and yards. While the effects of a few trees on a single parcel generally are noted anecdotally only, when large stands of trees are present, it becomes possible to analyze them in association with the surrounding population to understand what sort of population–environment dynamic might (or might not) be at work in a broader geographical region.

The focus of this study (Smith et al., 2014) was to explore the linkage between tree canopy and health, based on state community health expertise, in residential neighborhoods of Ann Arbor, and use that linkage as a guide to targeting neighborhoods in which to concentrate the planting of new city trees (1000 per year).

Health Impact Assessment: Strategy

The HIA examined the potential future health and psychosocial benefits associated with targeting tree planting in residential areas of Ann Arbor with lower tree canopy (less than 30%) and populations likely to be more vulnerable to extreme heat events based on an inferred increase in health vulnerability. This HIA was intended to inform the tree-planting strategy of the City of Ann Arbor Urban and Community Forestry Management Plan, by recommending priority neighborhoods for tree plantings. In June 2012, an advisory committee of community members, academic experts, and local government staff met to begin the project as a workgroup. The key set of steps in their general strategy (Human Impact Partners, 2017), or story, is shown in the graphic in Figure 19.1. That simple, conceptual graphic is followed by a more detailed graphic (Figure 19.2) that exhibits details of the steps and paths of impact that would be expected to evolve through increasing tree canopy over time.

Reactive Health Conditions

Based on available data (Smith et al., 2014), the health conditions most likely to react directly in response to tree cover were identified by the workgroup. In general, these are conditions that might arise in association with extreme

Figure 19.1 *General project strategy, summarized as a graphic guide.*

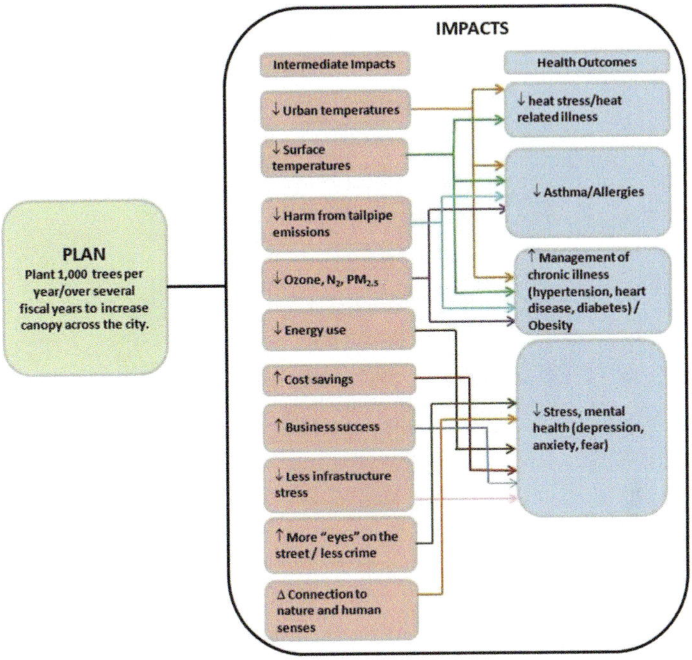

Figure 19.2 *Tree canopy impacts, direct and indirect. (Source: Smith et al., 2014.)*

heat or other forms of environmental stress: for example, in hot weather, folks may stay indoors more and exercise less, may not be able to afford air conditioning, and may experience less cardiovascular fitness and more mental stress. Or, if they do exercise in extreme heat, they may suffer health consequences.

The targeted conditions are shown in Figure 19.3.

Georeferencing of Data: Results by Neighborhood

When data sets associated with each of the six health conditions were compiled and associated with each residential neighborhood, and then mapped in association with tree canopy less than the city-wide average, the map in Figure 19.4 was the result. Here, the GIS software was critical in making associations between two different large data sets: tree canopy and inferred health vulnerabilities.

To see how this composite map was formed from a sequence of maps, an Esri cascading story map disentangles these data in a compact manner. All of the component maps appear, when viewed online, within a single frame on the computer screen, with associated text scrolling in and out as

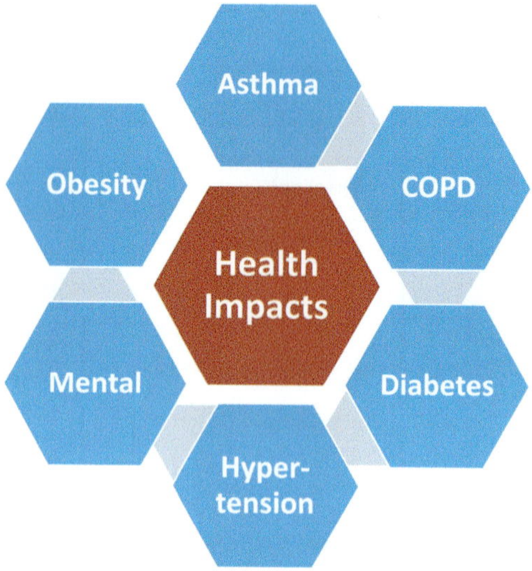

Figure 19.3 *Targeted conditions.*

appropriate. Scroll down and another map comes in; keep scrolling and the text associated with that map comes in. Figure 19.5 presents a sample of screen captures to illustrate the structure and content of that story map. In the full map, there are six maps that correspond to the six risk factors identified earlier, plus two additional maps showing neighborhoods with multiple risk factors. A QR code in that figure takes the reader to the full, interactive, cascading story map.

Groundtruthing Your Maps

As we evaluated the resulting maps that highlight the areas with likely disease vulnerabilities, we noticed that some were associated with student neighborhoods. In this case, our vulnerability index that takes into account income as a factor lets us draw maps of areas that appear poor because large numbers of students are included (students are included in the census) but are often not as poor as their census data appear to be. It is important to review the associations generated in any spatial analysis with locals to gain any special insight for the area.

Association and Causation?

The map in Figure 19.4 is one of the final frames in the atlas, summarizing the previous six maps focused on health vulnerabilities. In that map, notice that the areas that are light yellow are not directly involved in the analysis. Much

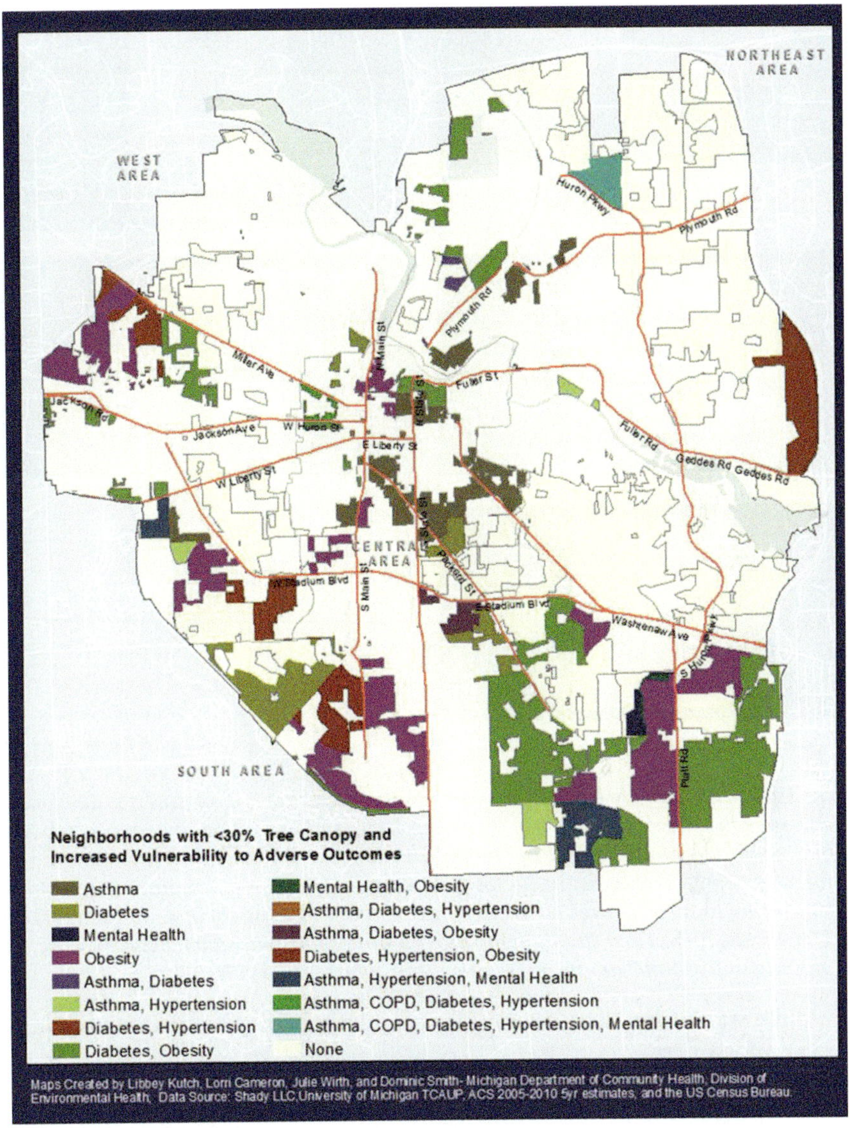

Figure 19.4 *Neighborhoods with lower tree canopy cover and inferred higher health risks.*

of that region is in fact covered by residential areas, but residential areas that have greater than average tree cover. Only those areas with less than average tree cover were highlighted on the map, while those with greater than average tree cover were not; thus, we do not know if areas with high tree cover suffer from similar health vulnerabilities.

This map suggests at least two directions for future general research:

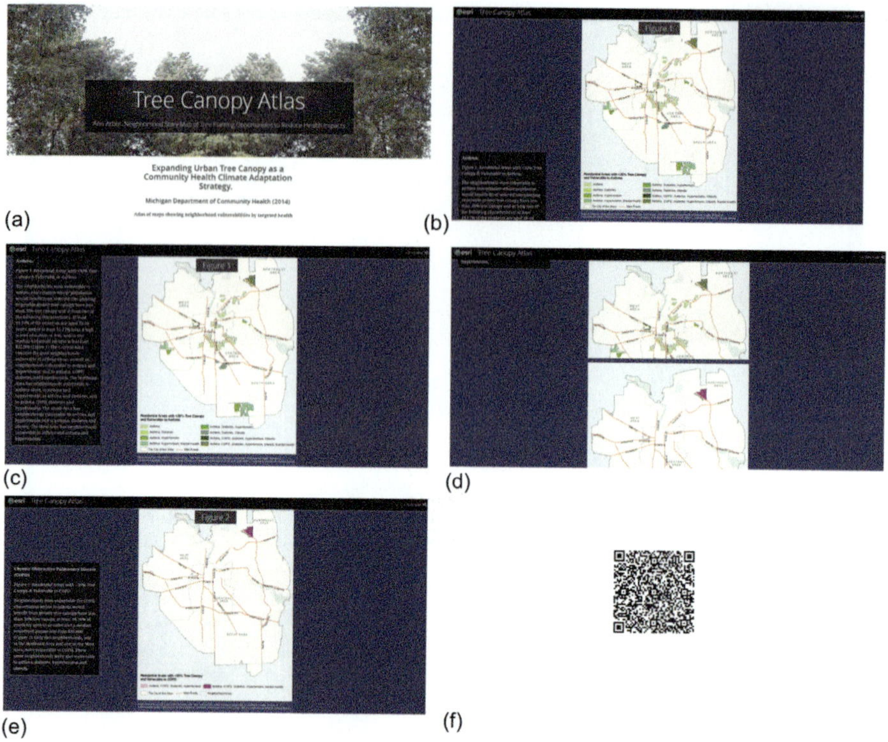

Figure 19.5 *(a) Shows the cover of the story map. Scroll down a bit and a map comes in, with some associated text in (b). Keep scrolling and more text appears in (c). Notice that the underlying map did not move; only the text moved. In (d), after more scrolling, a new map comes in. In (e) the new map fills the screen, replacing the old one; continuing to scroll brings in the new text without moving the new map. In (f) there is a QR code linked to a site showing the full story map (https://www.arcgis.com/apps/Cascade/index.html?appid=a4c44c18d91f4f2bbb2eb046357e7078).*

- Make similar health vulnerability calculations for all residential neighborhoods first.
  - Then, sort the data sets by tree cover.
  - Analyze the results and map them.
- Compare and contrast maps generated as follows:
  - Low tree cover and low health vulnerability.
  - Low tree cover and high health vulnerability.
  - High tree cover and low health vulnerability.
  - High tree cover and high health vulnerability.
  - Analyze the results.
- Compare and contrast the analysis using low-income families instead of low-income individuals to see if the "poor student" bias is reduced.

Reflections on One Neighborhood Not Singled Out by the Map

Now, the mapping and analysis may not be as straightforward as suggested by simple logic. For example, one neighborhood on the map, well known by Arlinghaus and known to have high tree cover, is a condominium association of 50 residences. Although there is no age requirement to live in that area, most of the individuals are elderly. One would expect neighborhoods such as this to stand out on a map of high tree cover and high health vulnerabilities. However, here the health vulnerabilities may have little to do with tree cover and more to do with the advancing chronological age of the residents. This neighborhood also has owners or renters with incomes that exceed the index criteria looking for lower-income populations.

In terms of tree canopy, recent devastation by the emerald ash borer in Ann Arbor has caused a substantial loss of ash trees. This particular condominium complex has about 17 acres of land, much of which is forested. Over the past few years, that complex has lost over 170 ash trees. Should they, or should they not, expect alterations in the health of residents due to reduced tree canopy? Over how long a period might change take place—20 to 25 years? Perhaps that is related to how long the various diseases take to develop, and how long it takes for new trees to develop to an extent where they offer significant shade. Is there mere coincidence in any association or is there merit in looking for causation? This is a different direction for further research, especially if person-level data could be obtained and tracked over time within the spatial and environmental context of changing canopy cover. The emerald ash borer event is an unfortunate but natural experiment opportunity.

References

Bergquist, P., Hadzick, Z., Kullgren, J., Matson, L., and Perron, J. 2012. Urban climate change adaptation: Case studies in Ann Arbor and Grand Rapids, Michigan. A project submitted in partial fulfillment of the requirements for the degree of Master of Science (Natural Resources and Environment) at the University of Michigan April 2012 Faculty advisors: Dr. Maria Carmen Lemos (Chair), Dr. Arun Agrawal. https://deepblue.lib.umich.edu/handle/2027.42/90864.

Gill, S., Handley, J.F., Ennos, R., and Pauleit, S. 2007. Adapting cities for climate change: The role of the green infrastructure. *Built Environment* 33, 115–133.

Human Impact Partners. 2017. HIP homepage. Retrieved from http://www.humanimpact.org.

Smith, D., Wirth, J., Cameron, L., Kutch, L., Skorokhod, V., and Stanbury, M. (Project Staff). 2014. Expanding Urban Tree Canopy as a Community Health Climate Adaptation Strategy. Michigan Department of Community Health: Ann Arbor, MI.

20

Web Maps as Story Telling

Joseph J. Kerski

Figure 20.0 Spatial word cloud summary, based on word frequency, made using Wordle.

Hands-On Activities

The following hands-on activities invite you, the reader, to build your own story maps. While following these procedures, think about the stories that you would like to tell. You can use the skills you gain in the hands-on activities to turn your own stories into multimedia story maps. These activities require a log-in to ArcGIS Online, from Esri. If you do not have an account, check with your organization—your university, your company, your government agency, your school, your non-profit organization. It is likely that your organization will already have Esri technology, including ArcGIS Online, that is the basis for story maps. If your organization does not have Esri technology, then go to http://developers.arcgis.com and scroll to the location where you can sign up for a free account. You can use this account to create, save, and share story maps.

Creating a Map Tour

The following procedures lead to the creation of a sustainable campus story map for the Esri headquarters campus in Redlands, California. The story that will be told with this map is that the founder of Esri, Jack Dangermond, is a landscape architect, and having grown up running his family tree nursery, has always loved living things. But southern California is an arid environment, and to keep the 3000 trees and plants of a wide variety of species on the campus alive in a more sustainable way, a holding pond north of the city was created. From the pond, non-potable water was brought to the campus in a system of pipes. The system worked so well that the excess water was given to the City of Redlands to water their trees in parks and median strips of boulevards. Another part of the sustainable campus story map is the solar panels on the roofs of the buildings, the bicycle storage, and the ride sharing programs. Many university campuses have sustainability initiatives, and this type of story map might work for those initiatives as well.

To create this map tour story map, go to the story maps apps page: http://storymaps.arcgis.com > Apps > Create Story > Map Tour > Sign in to your ArcGIS Online account. Next, go to the point where the system asks, "Where are your images?" > Select Google+ > Use the author's email address, jkerski@esri.com > Scroll to the folder "Esri Sustainable Campus" > Import. These images were taken with a smartphone with the location services on, so that they are in their correct position. If your images are not in their correct position, because they were taken inside a building, close to a canyon wall or building, or with a device that could not geocode the images, you can drag the photo positions on the map to their correct position.

Next, add additional captions as you see fit. Change the base map to "Imagery with Labels." Practice re-arranging the order of the images using the Organize tool. Save your map as "Esri Sustainable Campus" along with your initials or some other code so that you will be able to find it in the future. Within a few minutes, you now have a story map.

Dig deeper now and add data to your map. First, change one of the images to a video, as follows: Go to any photo > Change media > Video > Use the URL https://www.youtube.com/geographyuberalles and > Search the author's "geographyuberalles" YouTube channel for the following video: "A Tour of the Esri Campus in Redlands California." In YouTube, under the video, use Share > Embed > Copy and paste the embed code https://www.youtube.com/embed/TXKxi7jWwvc into the URL in your story map > Apply > Test your map and make sure the video is viewable.

Next, add a GPS track to the map. First, access the GPX file of the track on the Esri campus on https://www.josephkerski.com/resources/teaching and download this file to your device. This is an XML file of a GPS track that the author collected on the Esri campus. Go to http://storymaps.arcgis.com > My Stories > Find your Esri Sustainable Campus map > Go to Maps > Edit Map

> Add > Add Layer from File > Add your GPX > Find your local GPX file and add it > Then, click on and expand the layer in your map table of contents > Go to Trackpoints > Change Style > Use a yellow 8 pt.-sized symbol for the track points. Do something similar for the track line: Change style to yellow > Increase the width a bit > Save your map.

Next, go back to My Stories > Open your story map. Make sure the GPX track appears in your map. Go to ArcGIS Online with your log-in credentials (www.arcgis.com). Go to Content > Create a folder named "Storymap Esri Campus" > Move your story map and app to your new folder. Your map will look similar to that in Figure 20.1.

The next hands-on activity invites you to create another map tour, this time of locations that are not local but national in scope: Trees of Australia. Go to http://storymaps.arcgis.com > Apps (as you did before) > Create Story > Map Tour > Log in. Where the system asks, "Where are your images?" this time indicate: Flickr > Search the author's account "joseph_kerski" (with an underscore) > Find and select the trees of Australia folder > Import. Observe your map. Change the base map to an image base map. Consider adding other map layers that would help you understand the distribution of these trees in Australia, such as precipitation, ecoregions, or land cover. Again, you have created a story map in just a few minutes.

Next, add more descriptive captions. Change the base map to Terrain with Labels. Practice re-arranging the order of your images using Organize. Save your map as "Trees of Australia [your initials or some other code]" and using Contents in ArcGIS Online. Why use a code? That way, as you develop more maps, your codes will help you find your own content more readily. Move your map to an appropriate folder that you create, to help your work be as organized as possible.

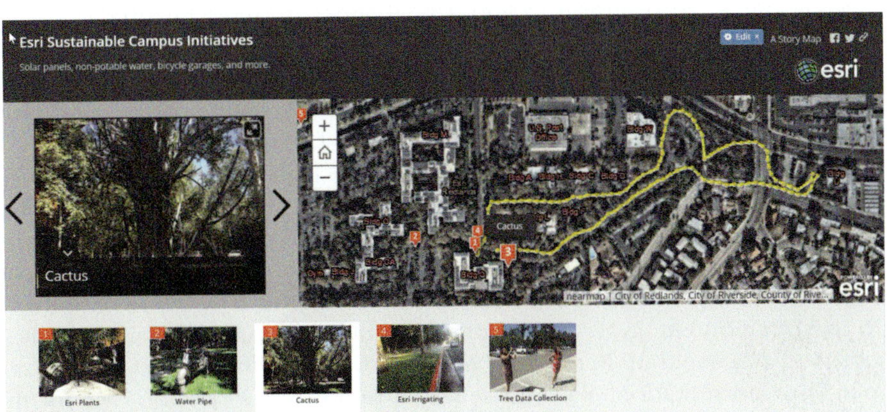

Figure 20.1 *Story map of the Esri sustainable campus initiatives.*

Creating a Story Map from a Web Map

Recall that story maps can be created from web maps in ArcGIS Online, because story maps use the same Web GIS platform, ArcGIS Online. The following activity will ask you to create a swipe story map with different content on the left versus the right side of the swipe—median age and median income (http://denverro.maps.arcgis.com/home/webmap/viewer.html?webmap=53f7c91aa02643c087e83a53edf87545). Or, start in www.arcgis.com (ArcGIS Online) and search median age and income "owner:jjkerski" with this exact syntax (colon after owner and no spaces). Make sure the median income appears at the top of the table of contents and is turned on, and your median age is underneath and is turned on. Next, use Save > Share > Create a Web App. Next, use Build a Story Map > Select Story Map Swipe and Spyglass. Give your map an appropriate title (include "Swipe" in your title to make it easier to find later, along with an alphanumeric code as you did with the previous activity) and provide some tags and metadata. In the builder tools that appear next, use Swipe Style Vertical Bar > Swipe Type: Select layer to swipe: 2014 Median Income > Take the defaults on App layout > Pop-up: Left Map header title: Median Age 2014. Right Map header title: Median Income 2014 > Open the app and check it. It should look similar to Figure 20.2.

Creating a Map Presentation

Recall the map presentation mentioned earlier. These are web map 'slides' that feature interactive maps and text titles. Build one in the following activity: First, in ArcGIS Online, go to your Content > Open your Median Age and Income map in the map viewer that you used in the previous activity. In the upper right, select Create Presentation (Figure 20.3).

Click on the + (add) button to add a new slide. In Slide #1, title this slide "Exploring Median Age and Home Value." Because scale matters, set the scale to current. Next, add another slide, Slide #2. Zoom so that you can see all of California by county. Set the scale to current. Add a title, "California Median Age." Add Slide #3. Zoom to San Diego County, in the southwestern corner of the state. Turn on Median Income > Title your slide "San Diego County Median Income." Set the scale to current. Click on a census tract > Check "Include open pop-up in presentation." Select Save. Next, play and check your presentation. Make sure that you are able to navigate between slides. The presentation should look similar to Figure 20.4 (Slide #3 shown). Go to your Content. Note that your presentation is part of your map; it is not saved as a separate item. From now on you can either open your map or open your presentation. You can send your end user a specific URL for the presentation only.

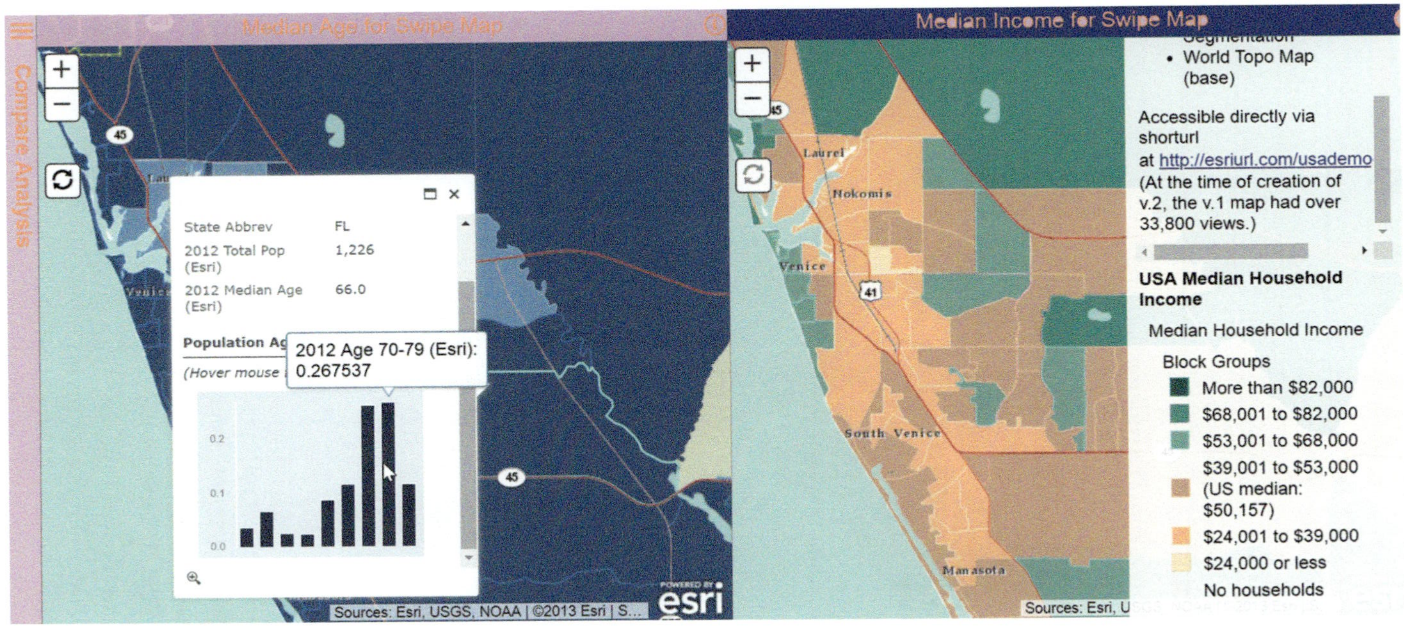

Figure 20.2 *Swipe story map of median age and median income.*

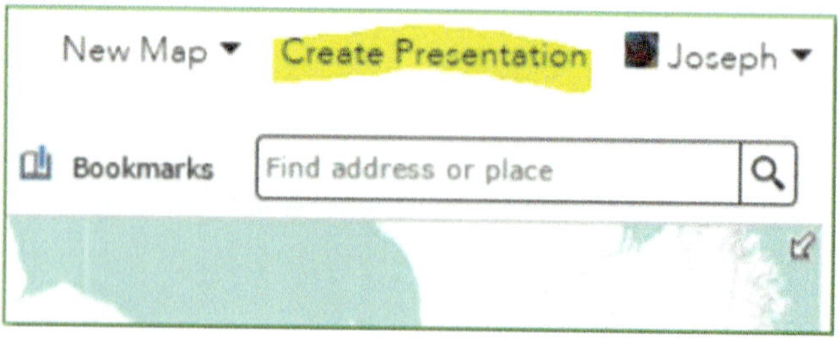

Figure 20.3 *Creating a map presentation.*

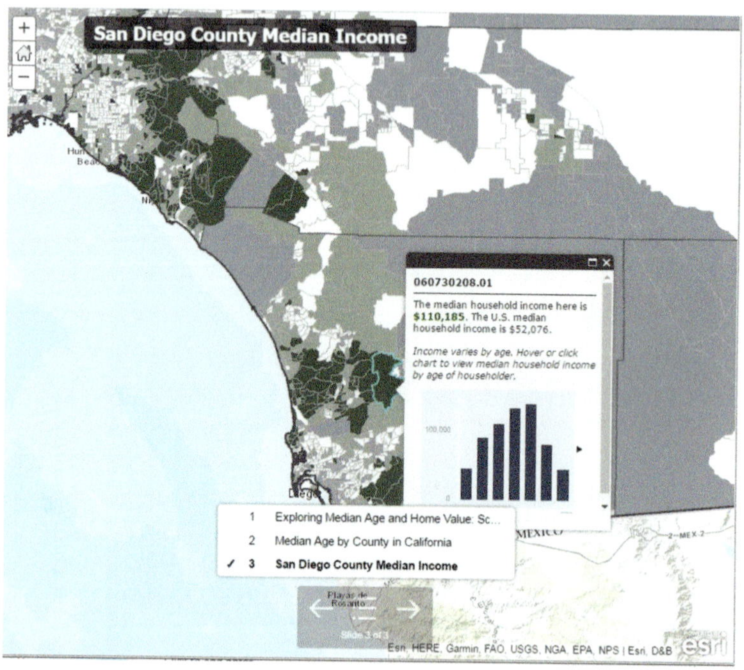

Figure 20.4 *Map presentation of median age and median income, slide 3.*

Creating Map Notes

Map notes are another simple yet powerful means of communicating. Map notes "pop up" when a point, line, or polygon is clicked, revealing information in the form of hyperlinks, text, images, or graphs. This activity guides you through creating map notes.

Open your Median Age and Income map in the ArcGIS Online map viewer. Make the population density layer invisible and the Median Home Value layer visible. Using the search box, find the latitude and longitude coordinates –117.168484, 32.706307. The result should locate you on the San Diego Bay shore. In the resulting find box, select Add to Map Notes. Next, edit the

Map Note. In the description, enter "View of San Diego Bay." For the image URL, enter http://www.josephkerski.com/wp-content/uploads/2012/07/sunset_sailboat21-1000x664.jpg. This photograph, as with other map notes that contain photographs, needs to be one that can stand alone on the web; in other words, it needs to be opened as a separate web page and not wrapped in an animation or frame. Next, select Edit to close your editing session. Test your pop-up. It should look similar to that in Figure 20.5. You used the image URL in this pop-up. One more URL is possible in a pop-up. This additional URL is the web page that will be accessed when the user clicks on the image. This could be a convenient way to direct your map users to specific content, such as your research project's web page or a web page that focuses on the type of feature in your map note, such as the tree type or an issue you are investigating, such as an urban greenway, dangerous intersections, or water quality.

Creating a Map Journal

In this activity, build a map journal with a sustainable agriculture theme. First, access the story maps page on http://storymaps.arcgis.com. At the top of the page, click Apps > Sign in to ArcGIS Online if you are not already signed in. Select the Story Map Journal. Click Build. Choose the Layout > Side Panel > Start. Enter a title: "Feeding the Planet by [your initials]." Configure content: Click Image > Flickr. Type in the user name "mapjournal." Click Load albums. Click album "Workshop (10)" (you can search for it using Control-F on the page; there are many folder albums to choose from). Choose any photo that interests you from the folder Workshop (10).

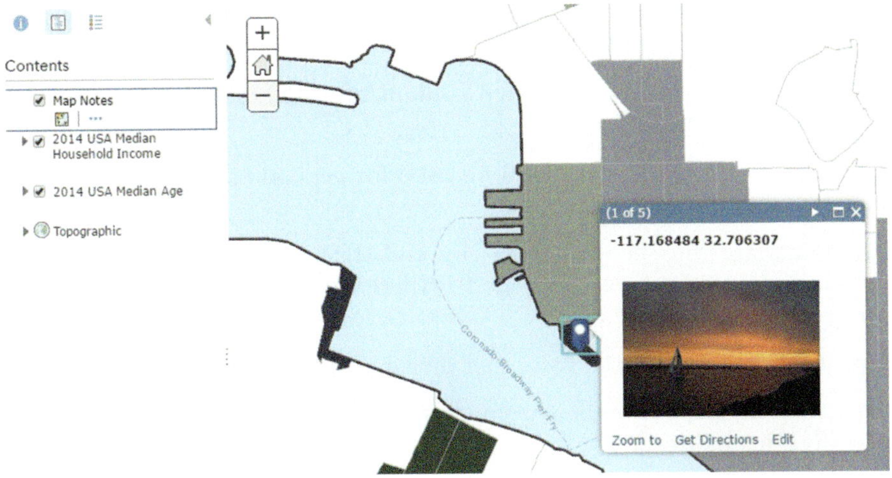

Figure 20.5 *Map note with multimedia.*

Confirm that Fill is checked > Click Next. Add text (e.g., "Sustainable Agriculture") into the text block. Next, click Add. Now, your Home section is done. Next, click Add Section > Enter title "Current Crop Yields." The first section contained an image. This section will contain a map. Use Content: Click Map > Click Select a web map then > Select a map. Search for the string "StoryMapWorkshop feeding" (note the lack of spaces in the first phrase, spaces and upper and lower case are important in searching) in ArcGIS Online. When found, click its thumbnail. Next, in Content, click Custom Configuration > Click boxes next to the following layers: Current crop yields notes, Current crop yields, and the Elevation hillshade. Click Save Map Content > Next > Add text > Click the camera icon to add a photo from Flickr or elsewhere.

Creating a Tabbed Series Story Map

In this activity, you will create a tabbed series story map. The theme of the map will be to study the geographic distribution of three divisions of junior colleges who sent their baseball teams to the Junior College World Series (JUCO) in 2015. Start with ArcGIS Online (www.arcgis.com) and sign in if you need to. Open the web map of the USA Tapestry Segmentation: http://www.arcgis.com/home/webmap/viewer.html?webmap=b8480d979e554729a25304afc829a727.

Tapestry Segmentation is Esri's geodemographic market segmentation system. It classifies US neighborhoods into 65 segments based on their socioeconomic and demographic composition. For a broader view of markets, segments are grouped into 12 LifeMode Summary Groups that reflect lifestyle or life stage of the households in each area. Examine the patterns at different scales: state, county, census tract, and block group.

Use the Add button and select "Add data from web." Then, add data for the baseball teams from the following spreadsheet that is online as a comma-separated value (CSV) file:

https://www.josephkerski.com/wp-content/uploads/2018/11/mesa_county_juco_div1_teams-1.csv

Indicate that the field "State" will be used for geocoding by state and "City" for geocoding the city (Figure 20.6).

The points are thus geocoded on city and state only (no street address or latitude–longitude values are needed). When mapping the points, change the attribute to "Champion."

Repeat this process for the two other divisions of baseball teams:

https://www.josephkerski.com/wp-content/uploads/2018/11/mesa_county_juco_div2_teams-1.csv

https://www.josephkerski.com/wp-content/uploads/2018/11/mesa_county_juco_div3_teams-1.csv

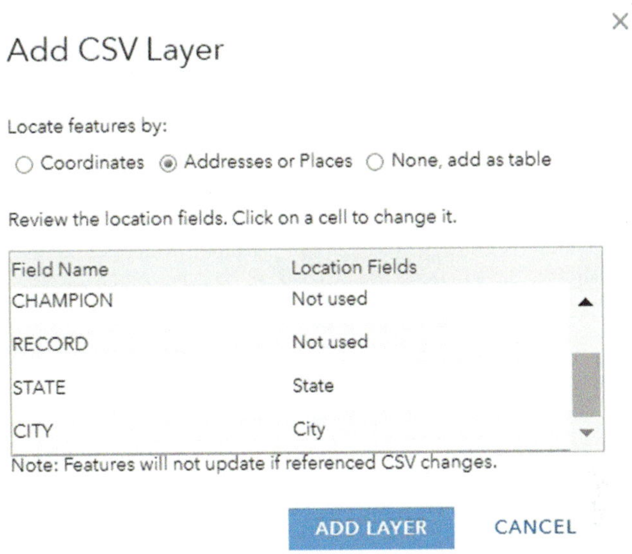

Figure 20.6 *Indicating fields that will be used for geocoding while building story maps.*

Use Change Style and symbolize each of the three layers as "Show location only" but symbolize each of the three layers uniquely so the three different divisions can be distinguished. Note the patterns or lack of patterns for the three divisions and compare and contrast them. Save the map and share it. Create a new web mapping application > Select story map > Select series story map. Add tabs for the tapestry and for each of the JUCO divisions, so that the final map contains three tabs. For the Division 1 tab, customize the map to show only the tapestry layer and the Division 1 layer. For the Division 2 tab, customize the map to show only the tapestry layer and the Division 2 layer. For the Division 3 tab, customize the map to show only the tapestry layer and the Division 3 layer. Save the story map. Add at least one link and at least one photograph or video to this map. Change the default Esri logo to a custom logo. When the user clicks on the logo, direct the user to a suitable URL, such as that of the National Junior College Athletic Association (http://www.njcaa.org/landing/index). Change two of the other default settings of your choice to further explore the customization possible. Save the map, and if you would like to share it, share only with your organization. Your map should look similar to that in Figure 20.7.

Digging Deeper into Story Maps: A Lakota Language Story Map

My colleague James Rattling Leaf and I (the author of this chapter) created a story map on the Lakota language (http://denverro.maps.arcgis.com/apps/MapTour/?appid=60ac74d36ae34ce181e88fbeeeb56831).

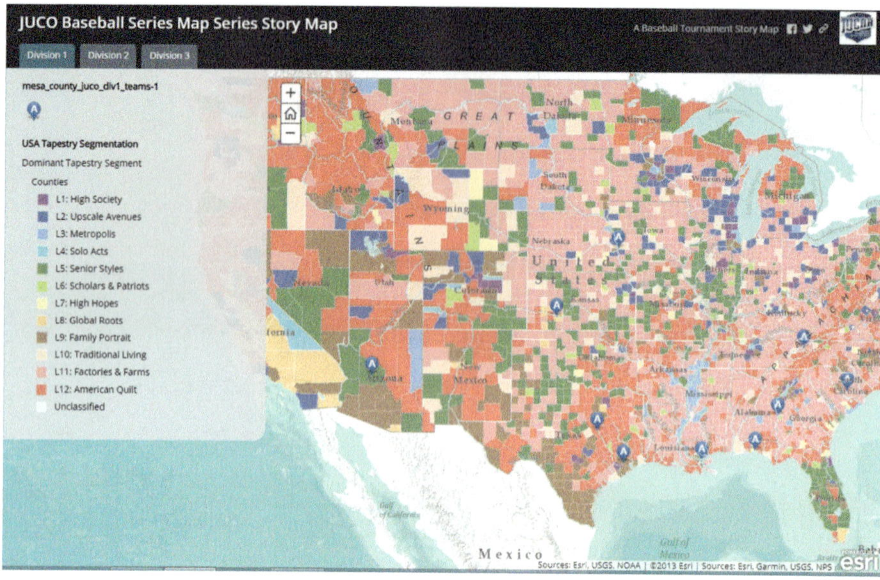

Figure 20.7 *Resulting story map from JUCO baseball tournament, Division 1, with tapestry demographics and lifestyle variables.*

Our reasons for doing so were several. We wanted to promote the use of story maps and geographic information systems (GIS), and provide an example story map that educators and students could easily create on a topic they were interested in. For this map, I interviewed James, saved the audio, broke up the audio into segments, one for each word or phrase, and stored the audio files on a website. In the story map captions, I pointed to the stored audio file with simple html code; html 5 took care of configuring the player to appear without any additional coding on my part.

More importantly, we have long been interested in and collaborated on projects involving education, maps, and GIS, and wanted to illustrate how the story maps platform can be used to learn and teach about Native languages, beginning with Lakota. When a person accesses the story map and steps through its contents, that person can hear audio of a dozen words that are in both Lakota and English, a photograph of each spoken feature, and what that feature looks like on a satellite image (Figure 20.8).

By coupling visual cues with audio, the goal for this map project was to inspire others living on the Lakota lands, those working with language projects such as Recovering Voices (https://recoveringvoices.si.edu), at the WoLakota Project (https://www.wolakotaproject.org), at the Language Conservancy (https://www.languageconservancy.org), and others, to take these ideas and do even more with story maps. For example, one could embed these story maps in web pages; one could add video to the maps (as we illustrated with the word "lake"), one could create different types of story maps, and much more.

Figure 20.8 *Lakota language story map, as a map tour.*

For learning about language, place, biology, history, geography, and many other themes, integrating audio and video with maps is becoming a powerful and yet easy-to-understand medium.

Second, we were interested in the issue for reasons deeper than our affinity for languages, geography, and GIS. As noted on Lakhota.org:

> Native languages in the United States are in the throes of a prolonged and deadly crisis. For the past 400 years, Native Peoples and their languages have been steadily and undeniably disappearing. Though the historical fate of Native Peoples has been reluctantly acknowledged, less is publicly known about the associated fate of their languages.

And furthermore,

> Lakota is dangerously close to extinction. Recent linguistic surveys and anecdotal evidence reveal that Lakota speakers of all abilities, on and around the reservations of North Dakota and South Dakota, amounted to fewer than 6,000 persons, representing just 14% of the total Lakota population. Today, the average Lakota speaker is near 65 years old.

Furthermore, geography, place, location, and culture are reflected in the Lakota and other Native languages, making story maps an excellent tool for teaching and learning. For example, according to Lakhota.org, "Nature is used as the primary source for the metaphor models." Furthermore,

> Lakota is also very good at emphasizing the finer attributes of travel. A person can be considered to be coming or going to or from specific places in many levels detail. Lakota greetings themselves reflect this tendency, where in English "welcome" is literally Lakota—"Good that you came," And "goodbye," is "Travel well." The language also closely linked the land to the people through geographical names and stories. [...] A word like *woímnayankel* expresses notions of awe, humility, and interconnectedness. A Lakota speaker might use this when describing the experience of the northern lights (aurora borealis). The word expresses the humility that a person feels when confronted by the awesomeness of nature while also feeling intimately connected with it.

Consider how you might you use story maps, and the ideas presented through this Lakota language story map, in your own work.

Sounds of Planet Earth Story Map

Planet Earth is a noisy place. Some sounds are made by objects on, above, or below the natural landscape, and others by things that humans have constructed on that landscape. I (the author of this chapter) created a "Sounds of Planet Earth" (https://denverro.maps.arcgis.com/apps/MapJournal/index.html?appid=7b3e4bc7d340445fa543e9b1f3b30280) story map with educational and technical goals in mind (Figure 20.9). One educational goal was for students

Figure 20.9 *Sounds of Planet Earth story map.*

to be able to use their sense of hearing to test their knowledge about 100 earth sounds. After listening, they can take a quiz for each sound to test their answer on what each sound is. Some are easy, while others are intentionally very challenging.

Each point on the map is associated with a sound. Users can navigate to each sound through the map interface or use the navigation bar on the left side of the map to move to each sound in numeric order, or feel free to skip to specific sounds. Once the user chooses a sound, a player appears, allowing him or her to listen to it. A link appears below each sound for the user to submit his or her guess as to what that sound is. The quiz is optional but is a fun, interesting way to engage with the content (Figure 20.10).

Each sound player shows the interactive web map next to it; for example, Sound #20. In this case, the map gives a good clue, such as below (Figure 20.11), but in other places around the Earth, the map provides very little additional aid.

Figure 20.12 shows the answer for Sound #20. In this case, a few answers that would be acceptable are provided, instead of just one answer.

This story map was created in the following manner: First, a spreadsheet was created with city, state, and country fields for each sound. My goals included the desire to have sounds of nature and sounds created by people, a diversity of different places around the world, to spark spatial thinking, and for them to be interesting. I used my own sounds because I own the content for those sounds and did not need to seek permission to use them.

Second, in ArcGIS Online, I added the spreadsheet that I had just created and created a feature service and a map from it. In the map, I used a custom sound speaker graphic created by a wonderful artist that I know for the point

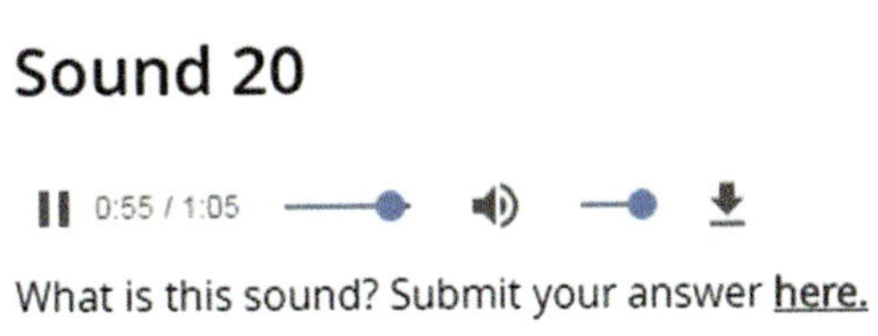

Figure 20.10 *Sound player in story map.*

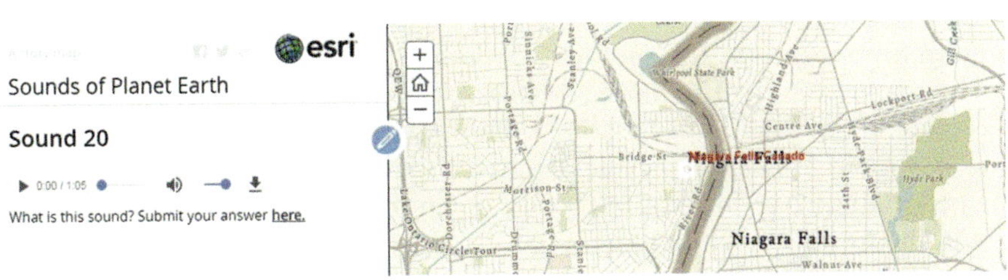

Figure 20.11 *Map and sound in a story map.*

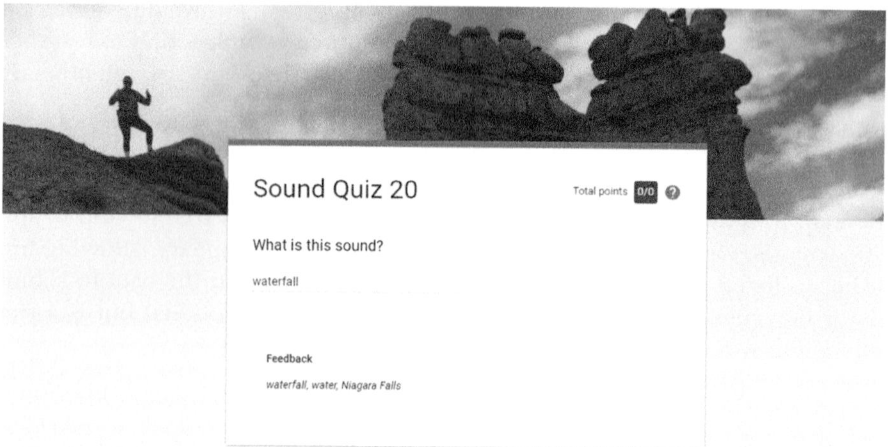

Figure 20.12 *Quiz embedded in a story map using a Google Form.*

symbol. Using the technique to access the new basemaps that I discussed here, I used the Modern Antique base map for this story map for something fun and different. I used the new functionality with Arcade expressions to label my features. I shared the map with everyone.

Third, I selected and downloaded 100 of my videos from my YouTube channel "geographyuberalles" and used Camtasia from TechSmith to separate out the video from the audio. Web utilities exist that allow a person to download audio files from YouTube, but they should only be used with extreme caution; most of them contain malware and worse. Camtasia or another video editing tool will work very well for the task of separating audio and video, and is much safer; plus, one can trim the sound clips. Why did I download the videos in the first place? I originally wanted to point directly to my own YouTube audio from my story map, but after several tries I could not get this method to work. So, plan B was implemented by downloading, stripping out the audio, and uploading, as is explained next.

Fourth, I converted 100 sound files from wav format from Camtasia to .mp3 format using Format Factory. Format Factory works well for conversion of audio and video files into different formats.

Fifth, I uploaded the 100 .mp3 sound files to a library on http://www.archive.org. Archive.org is the Internet Archive, a non-profit library of millions of free books, movies, software, music, websites, and more. For each file, I provided metadata and made them Creative Commons in licensing so that anyone can now use them. In a spreadsheet (separate from my spreadsheet that I used for my geocoding) I recorded the URL for each .mp3 file so I could point to the sound file in my story map.

Sixth, I created 100 quizzes and answer keys using Google Forms. I then viewed each quiz, recording the URL in my working spreadsheet. Seventh, I used the Map

```
<p>

<audio controls=""><source src="https://archive.org/download/niagara_falls_201707/niagara_falls.mp3

" type="audio/mpeg" /> does not support?</audio>

</p>

<p>What is this sound? Submit your answer <a
href="https://docs.google.com/forms/d/e/1FAIpQLSdBrJ1I6MF2CFu004hiPgcMGq8NPT3I4ZWln491YBN
zcXeyoQ/viewform

" target="_blank">here.</a></p>
```

Figure 20.13 *Code for embedding sound into a story map.*

Journal story map app to build my sound story map, pointing to the map that I created in step 2. For each sound, I set the appropriate map scale and extent. For each sound, I edited the html, pointing to my sound file and to the quiz.

When done, each sound was tested to make sure it was working, and that the correct quiz and answer key was associated with each sound.

One of my blocks of code is shown in Figure 20.13, with the first string representing the sound file and the second string representing the quiz. So, it shows that if an instructor encourages his or her students to get comfortable with a little bit of coding, it can reap great rewards.

We encourage readers of this chapter to continue their own exploration by studying physical and cultural geography, getting out onto the landscape, and encouraging students to observe with all five senses. Then, create a story map like the one described here with your own sounds, photographs, text, and/or videos. These techniques could be copied, or you could try other methods of creating a sound-based story map.

Other story maps incorporate sound in rich and varied ways. These include Alan Lomax's documentation of the music of the South (http://ipdemos.maps.arcgis.com/apps/MapJournal/index.html?appid=0f79dcd3b6504be5b39102113 8d3b7f4), a concert tour by the Grateful Dead (http://www.arcgis.com/apps/MapTour/index.html?appid=fd9bf89b12d54ee8958fe41fca83f46c), and more Grateful Dead from 1977 (http://modelcounty.maps.arcgis.com/apps/MapJournal/index.html?appid=5b63b952049a4868b301878e686d0e6a), some traditional Christmas holiday music from around the world (http://storymaps.esri.com/stories/2013/holiday-music), and the Sound of Music filming locations (http://zgis.maps.arcgis.com/apps/MapTour/index.html?appid=0c31f863 93d8405daf4c2d95acd7e947).

Another story map incorporating a quiz is this map about ports around the world (http://pot.maps.arcgis.com/apps/MapSeries/index.html?appid=a84f271 4c7874715912fb4c6bd18ed56). This quiz uses Survey123, and illustrates that these tools (in this case, field tools and web mapping tools) can be combined in creative ways, because they are all using the same Web GIS platform. I created a "Name that Place" quiz, using the presentation mode discussed in this chapter, here: https://www.arcgis.com/apps/presentation/index.html?webmap=f95d562571d740a6840254ee53ae3024.

Using Images in Web Maps and Story Maps

Many options exist for using images in ArcGIS Online, including web mapping applications such as story maps. Choose a method that works best for your situation and needs. The most commonly used methods and photo archive services are described in a series of guidelines by the author of this chapter (https://community.esri.com/community/education/blog/2017/02/17/photo-guidelines-for-arcgis-online-maps-including-story-maps). These guidelines represent only a subset of the total number of methods that are valid. The guidelines describe the use of images in ArcGIS Online, Flickr, Microsoft OneDrive, Dropbox, Google Photos, and Google Drive. Each has its advantages and challenges.

Some map creators do not use any of these photo archive services, but instead favor the use of ArcGIS Online for the storage of their images. This method offers the convenience of storing maps and photos all in a single location, avoids the challenge of the rapidly changing interfaces and capabilities of these photo services, and gets around the issue that some educational institutes have of certain photo archiving sites being blocked. The author of this chapter prefers to use ArcGIS Online for his maps and map layers, and a separate photo archiving tool (e.g., Flickr) for images. Using these procedures takes advantage of the "best of both worlds"—using ArcGIS Online for the mapping and using the easy-to-access photo archive services for the images. But again, it is recommended that the reader choose the method that works best their own needs, situation, and institution.

Two rules of thumb are important when using images in web maps and story maps. First, ArcGIS Online and apps (including story maps) are continually evolving. The photo sharing tools are likewise continually evolving. Thus, these guidelines and procedures are subject to change. Second, to be successful with using photos in ArcGIS Online, the following points are important: One, make sure the photographs are your content that you have created yourself, or they are in Creative Commons or are not copyrighted, or else you have permission to use them. Two, make certain that the photos are shared with the public. Three, make sure that the photos are of modest size, such that they will not slow down the browser, but are not so small that they will appear grainy. Four, make sure that you can obtain a URL that can be opened in a separate tab in a web browser. If the images can be opened in a separate web browser tab, then they will work in ArcGIS Online.

At the time of this writing, Flickr offers the best platform for storing images. Flickr allows for albums to be easily created and named in intuitive ways, rather than naming them with today's date, as Google+ did before sunsetting in 2019. Flickr also allows for the image URL to be exposed and used, which is critical in using them in maps. Images will not work in maps generally if they are embedded in an animation or in some sort of web frame. If the image via its URL can be opened in a web browser tab, then it will work inside a web GIS or a story map. Opening images in a browser tab first is thus a good test to determine whether they will work inside a web map. Flickr also provides at least eight different URLs that correspond to different sizes, in pixels, of each image (such as 500 × 500 pixels). This URL-by-size feature is very useful for obtaining thumbnail URLs as well as for obtaining full-sized image URLs for story maps. This automatic by-size feature also eliminates the need to use an image editing program such as Photoshop for resizing the images.

Story Maps for Landscape Comparison

Many educators, researchers, students, and analysts regularly want to examine changes over space and time with imagery and GIS (Milson et al., 2012). Recently, 81 different dates of historical imagery for the past 5 years were placed inside ArcGIS via the World Imagery Wayback service (https://www.esri.com/arcgis-blog/products/arcgis-living-atlas/imagery/wayback-81-flavors-of-world-imagery). This imagery is accessible in ArcGIS, ArcMap, and ArcGIS Pro. The best place to start is the World Imagery Wayback app. This app, available through a web browser (https://livingatlas.arcgis.com/wayback) can be used by way of introduction in a university or community college course, or all by itself in a primary or secondary school. A fascinating and an incredible resource for examining land use and land cover change, the Wayback image service covers the entire *globe*. That means that educators and researchers can examine coastal erosion in England, deforestation in Indonesia, urban sprawl just about anywhere, reclamation of mine lands, changes in water levels in reservoirs, agricultural expansion in Saudi Arabia, glacial retreat in Alaska, and much more.

In addition, in keeping with one of the themes of this chapter (and the Spatial Reserves blog and book; https://spatialreserves.wordpress.com), be critical of the data in GIS. The Wayback app and imagery create useful 'teachable moments' in fostering critical thinking. First, the dates shown on the left-hand side of the app (Figure 20.14) represent the update of the Esri World Imagery service, fed by multiple sources, private and public, from local and global sources. Thus, the date shown does *not* mean that every location that can be examined on the image is current as of that date. For example, I verified this in my local area, where I observed construction as of June 2018, but this construction does not appear on the image. In addition, several other places I examined that were time stamped during the winter season in the Northern Hemisphere were clearly 'leaf-on' and taken the summer before. Therefore, I recommend becoming as familiar as possible with what you are working with. Despite these cautions, the imagery still represents an amazingly useful resource.

Figure 20.14 *Sample from the Wayback imagery data set for July 30, 2014 (top) and 4 years later, June 27, 2018 (bottom) for an area outside Denver, Colorado.*

How can the use of the Wayback image service be extended for education and research purposes? One way to do so is by creating a story map from the Wayback app. Doing this will thus enable the user to use all of the functions in ArcGIS along with the imagery, such as adding additional map layers (e.g., hydrography, land use, ecoregions), saving and sharing, using the measurement tools, and creating web mapping applications from the map. To create a story map, perform the following steps:

1. Go to the app (https://livingatlas.arcgis.com/wayback).
2. Navigate to an area of interest.
3. Check "Only updates with local changes" (Figure 20.15).
4. Click the cloud icon to "Add to cart" (Figure 20.15, at right).
5. Click the "Clear all" icon top-left to create a web map (Figure 20.15, at top left).
6. Save the web map.

The web map has been created. Open this web map. Now, layers can be added, including additional Wayback layers. To add the historical Wayback imagery to this existing web map, use Add Data and search in ArcGIS (do not search the Living Atlas), as follows (Figure 20.16).

The default sort order is by "relevance" but can be changed to sort by title or by oldest or newest. The resulting map shows three historical layers along with the current image as a base map (Figure 20.17).

Another way to dig deeper into change-over-space-and-time analysis with the Wayback image service is to create a swipe story map. A swipe map is a type of story map application that is quite suitable for examining change because it allows the map user to swipe across a map that has, in this study, images with two different dates. To create a swipe map, in ArcGIS, Share > Create a web mapping application > Choose Swipe map. Select one of the historical image layers for your swipe map, and make sure the base map is Imagery or Imagery with Labels. The swipe layer (the historical image) will appear on the right with the more recent image on the left.

Suppose, however, that a goal of a researcher or instructor is to have the left side be the older imagery, and the right side be the newer imagery. This is also possible. The swipe map template only allows the researcher to swipe one layer, which by default is the right side. Thus, to create a map with the

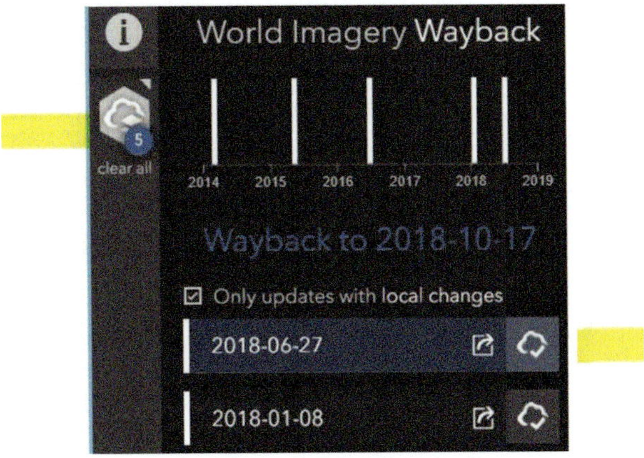

Figure 20.15 *Working with the Wayback historical satellite imagery.*

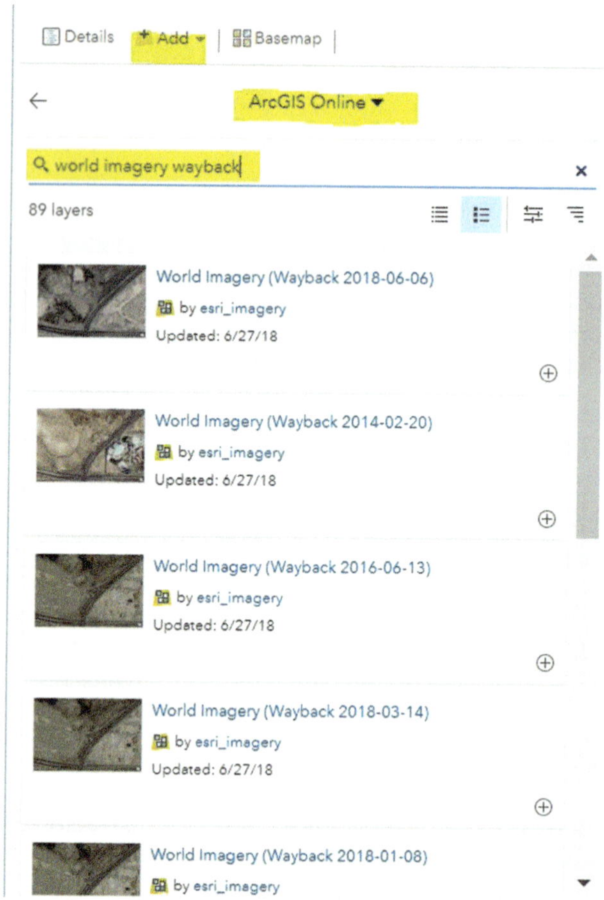

Figure 20.16 *Adding historical imagery from the Living Atlas of the World.*

older image on the left side, make the left side, the base map, a historical image rather than the default new imagery base map. To do this, go back to your ArcGIS map and select Add > Add from ArcGIS > enter "Wayback" > choose a historical image (in my case, I chose 2014) > Add as base map. Save your map. In the configuration panel for your story map, change the settings so that you are swiping one of your newer image layers. The results are shown in Figure 20.18, with the URL of http://denverro.maps.arcgis.com/apps/StorytellingSwipe/index.html?appid=67c6d5069feb42cca2de967781ead011.

Other possibilities exist for the use of the Wayback imagery, including using it in 3D scene for a historical perspective on the landscape, using them in a tabbed series story map, using them as a base for advanced analytics in ArcGIS Pro (e.g., the following blog post about bringing the data into Pro: https://www.esri.com/arcgis-blog/products/arcgis-living-atlas/imagery/wayback-server-connection-in-pro), and in many other ways.

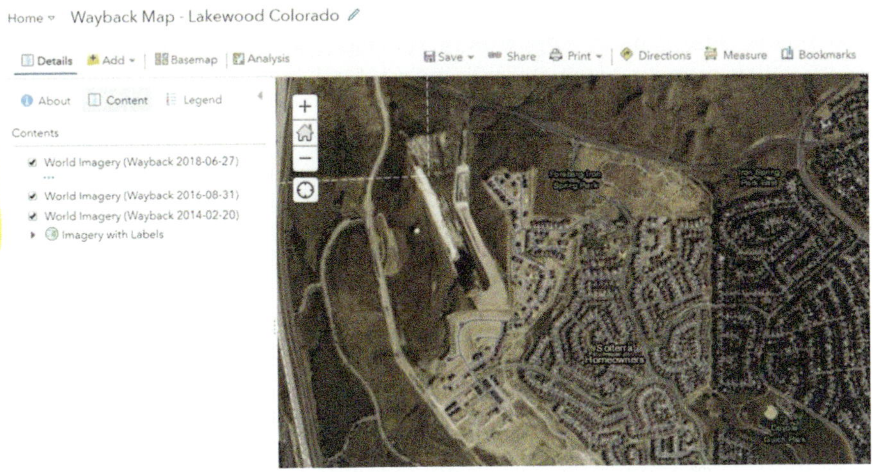

Figure 20.17 *Historical satellite imagery shown in ArcGIS Online.*

These interlocking tools and data are encouraging educators and others use story maps with each new data, functions, and tool in the rapidly evolving GIS environment.

Story Maps: The New Medium for Journals, Books, Multimedia, and Other Communication

Story maps are increasingly embedded into an ever-growing set of multimedia. Through the provision of i-frame embed code, any story map author

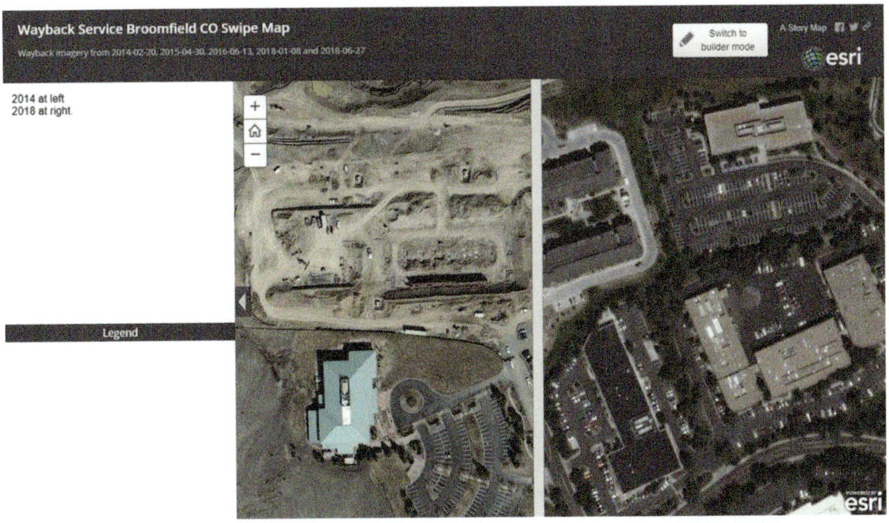

Figure 20.18 *Swipe story map with older (left) and newer satellite imagery (right).*

can embed the story maps into Sway or Prezi presentations, or standard web pages. The Alaska Department of Fish and Game, for example, created a story map to teach its citizens in Anchorage how to co-exist with bears (http://www.adfg.alaska.gov/index.cfm%3Fadfg%3Dlivingwithbears.anchorageurbanbearsstorymap). The story map is embedded into the state of Alaska's own web page. The Interesting Behaviors tab shows videos taken from cameras on bear collars, and as it shows bears swimming, taking care of their young, crossing streets, and other activities, it makes an effective instructional tool!

Story maps are being used in annual reports, to introduce people to an organization, to promote and market events, for project or business portfolios, for tutorials and lessons, for showing real estate, for online guidebooks, for constituent engagement in politics, for activism, for atlases, for binders, catalogs, briefings, presentations, newsletters, and web pages, in journalism—even in wedding invitations (Szukalski, 2018). One of the most exciting things about this list of ways story maps are used is that the bulk of the use is outside the GIS community. This shows that one does not have to be a GIS expert to create story maps, and it is an encouraging sign that people are beginning to embrace maps as more than reference documents. The volume of story maps created surpassed 2000 per day in late 2018 with no signs of abating.

As they grow in number and variety, story maps are also beginning to be placed into atlases or collections. An early example is the set of maps in the Atlas for a Changing Planet: http://storymaps.esri.com/stories/2015/atlas-for-a-changing-planet. Another example of a story maps collection is the Age of the Anthropocene collection of eight story maps, each containing several map layers and narrative, mentioned earlier. A recent example is a collection for the Digital Humanities (https://collections.storymaps.esri.com/humanities). As described by story maps founder and cartographer Allen Carroll (https://www.esri.com/arcgis-blog/products/story-maps/sharing-collaboration/story-maps-and-the-digital-humanities), digital humanities is the creative application of digital technology to humanities questions and data, the use of computational techniques in the humanities that would allow research that is otherwise impossible, and the democratization of knowledge through the application of digital technologies to the advancement of discourse in the humanities, broadly defined.

This collection includes a series of story maps called "New Brunswick Loyalist Journeys" by the University of New Brunswick Libraries. An introductory essay is followed by 'carto-biographies' of 10 'York County Loyalists' who fled the fledgling United States after the Declaration of Independence. Each biography links to footnotes, bibliographies, and illustrations, which are also presented in the form of story maps. Another story map in the collection was created by novelist Susan Straight, who compiled a list of 737 novels about the 'American Experience'. An avid reader, Susan locates each novel's setting, shows the map reader each book's dust jacket, and provides eloquent, single-sentence tributes for each volume. "Indiana Limestone," a story map by the Indiana Geological Survey, explores the cultural and historical dimensions of the extensive deposits of the limestone that underlie much of south-central Indiana. The story maps document the deposits and portray the quarry workers and stone carvers who have worked the stone for nearly 200 years.

In academia, many scholars are using story maps over PowerPoint and other traditional means as their primary presentation tool while teaching in the classroom. Some have inserted maps into their academic blogs, annual reports, proposal progress reports, theses, dissertations, and even peer-reviewed journal articles (Wright, 2018a). Progress is being made to make these story maps citable and also giving them the ability to have a digital object identifier (DOI).

Some have created story maps as an interactive supplement to their CVs or resumes (e.g., CVs from Joseph J. Kerski [https://denverro.maps.arcgis.com/apps/Cascade/index.html?appid=c84bb188001746d1a5ca43f83b366c66] and Amanda Huber [http://www.arcgis.com/apps/Cascade/index.html?appid=bc4a4fee3e88404a873277375bddf511]).

Story maps are also increasingly used as scientific and communication tools. In one example, the GeoDesign Summit, which brings together design professionals, GIS professionals, landscape architects, and those in GIS education and research from universities, I personally witnessed just about every presenter using a story map as the means of communication. In another extensive example, Dr. Dawn Wright describes how these maps are being used to communicate research results (Wright, 2018b) (https://community.esri.com/groups/story-maps-for-education/blog/2018/10/07/speaking-the-language-of-spatial-analysis-via-story-maps). The examples include analyses of automobile accidents, drought, floods, wildfires, energy from the ocean, land use change, ocean acidification, mortality in past conflicts, and many more subjects.

Story maps are beginning to be used as the platform for digital journals and books. Most textbook publishers began to turn their offerings into digital books 15 years ago and more recently into interactive 'courseware' that includes multimedia content, quizzes, and ways for readers to interact with each other. Several of these textbook publishers are beginning to move the maps in this e-courseware that are still by and large static images into interactive maps such as story maps. Wiley was one of the first publishers to do this, for their *Human Geography: People, Place, and Culture* book (Fouberg et al., 2015).

Some authors are bypassing established publishers and moving into story maps directly for publishing their work. For example, in 2017, a group of geography professors in Ethiopia used story maps for the entire contents

Figure 20.19 *Article in Focus on Geography that illustrates the pairings of stories and interactive maps that are becoming more common in journals, magazines, and books.*

of a new Ethiopian geography textbook. Another example is the Age of the Anthropocene collection (https://storymaps.esri.com/stories/2018/anthropocene-atlas/1-human-reach.html), which contains maps and narratives about population density, cities in 1950 and 2020, Earth at night, roads, air routes, shipping routes, undersea cables, and the human footprint. This collection could be effectively used as a textbook in a cultural geography course.

In the field of journal publishing, in 2016, the American Geographical Society relaunched its peer-reviewed journal *FOCUS on Geography* in a dynamic, interactive format. While this format does not use Esri story map technology per se, the look and feel of it is very similar, with a scrolling bar that allows the reader to move through the content and an interactive map (http://www.focusongeography.org), as shown in Figure 20.19. One example is an article about the pros and cons of paving the Linden–Lethem trail in Guyana, and the larger challenges and opportunities facing the country (Barton, 2017). As the reader navigates the 21 frames in the right-hand bar, he or she is presented with text, images, and videos, while on the left, the map automatically scrolls to the location on the Linden–Letham's 276 mi. (444 km) length where that part of the story takes place.

I predict that these developments are harbingers of an entire set of coming digital journals, digital magazines, and e-books that will based on story maps principles.

Story maps may be fairly simple and straightforward to create, but as reflections of broader trends in GIS, they are changing the very nature of GIS (Kerski, 2017, 2018). They are also changing the way that millions of people are interacting with maps and understanding their world.

References

Barton, K. E. 2017. Guyana's Linden to Lethem Road: A metaphor for conservation and development. *Focus on Geography* 61. http://focusongeography.org/publications/photoessays/guyana/index.html.

Carroll, A. 2018. Story maps and the digital humanities. ArcGIS blog, Esri, August 9. https://www.esri.com/arcgis-blog/products/story-maps/sharing-collaboration/story-maps-and-the-digital-humanities.

Fouberg, E. H., A. B. Murphy, and H. J. de Blij. 2015. *Human Geography: People, Place, and Culture*, 11th Edition. Wiley.

Kerski, J. J. 2017. Connecting citizen science, GIS, community partnerships, and education. Citizen Science GIS, blog. University of Central Florida. http://www.citizensciencegis.org/connecting-citizen-science-gis-community-partnerships-and-education-a-guest-blog-by-dr-joseph-kerski-of-esri.

Kerski, J. J. 2018. Why GIS in education matters. Geospatial World, blog, August 6. https://www.geospatialworld.net/blogs/why-gis-in-education-matters.

Milson, A. J., A. Demirci, and J. J. Kerski. 2012. *International Perspectives on Teaching and Learning with GIS in Secondary Schools*. Dordrecht, the Netherlands: Springer.

Szukalski, B. 2018. ArcGIS Blog, Esri, October 16. https://www.esri.com/arcgis-blog/products/story-maps/constituent-engagement/things-you-can-do-with-story-maps.

Wright, D. 2018a. Making story maps citable (e.g., with digital object identifiers). GeoNet Blog, Esri, October 8. https://www.esri.com/arcgis-blog/products/story-maps/sharing-collaboration/making-story-maps-citable-e-g-with-digital-object-identifiers.

Wright, D. 2018b. Speaking the language of spatial analysis and science via story maps. GeoNet Blog, Esri, October 7. https://community.esri.com/groups/story-maps-for-education/blog/2018/10/07/speaking-the-language-of-spatial-analysis-via-story-maps.

21

In Closing: Spatial Thinking, from Evolution to Revolution

Sandra L. Arlinghaus, Joseph J. Kerski, Ann Evans Larimore, and Matthew Naud

Evolution

From animaps, to Google Earth, to GEOMATs, to story maps, the reader has been led through a menu of intertwining methods of a given decade, all set within an environmental context of one sort or another.

Where do we imagine such interaction will lead in the future? Here are several bold views offered as possible glimpses of a broad future within which spatial thinking and other activities will take place. Will these glimpses be realized? Only time will tell. We offer them here not to see if we are accurate in forecasting but rather to encourage those who engage in spatial thinking to do so with an open and imaginative approach. The sky is not the limit; look beyond the conventional.

Computer software is a moving target. Spatial thinking should not be tied to technology; it should use technology appropriately. Big questions and issues require a bold approach in both teaching and research. We have various thoughts, dreams, and issues regarding the future. These thoughts are likely not original, but they are thoughts that a couple of us would like to share. We hope these viewpoints encourage readers to create their own sets of thoughts as well and then to act on their own rather than being dominated by those of an 'in' group of others or by technology itself.

Revolution

Arlinghaus:

- As medical science progresses, perhaps a 100-year lifespan will be the expected value for humans, and 150 years for a particularly long life. What are the implications of that idea in regard to global planning for that future?
- As robots and other technology bring us ways to acquire goods and services, and to communicate, we may find less need to cluster together physically (in cities) to gain economies of scale. If economies of scale diminish in importance, what are the implications for planning for all systems?
- When institutional education in bricks-and-mortar campuses fades out of the picture, will we be prepared to offer not only academic content, in a systematic and distributed manner, but also human/cultural experience to emerging generations that will help them interact successfully in a changing world that has undergone a cultural transformation?
- What will be the role of the institution of the library in the future? It is an enduring one, from the great library of Eratosthenes in Alexandria, throughout ensuing centuries. Will the library be the remaining vestige of contemporary educational institutions that moves forward with whatever a transformation might bring? If so, what plans should we make to try to protect it in various ways? What sorts of plans do we need to implement to back up the achievements of our human world?
- Altitudinal zonation is an old concept: as one climbs a mountain, he/she might first ascend through a zone (girdling the mountain) of deciduous trees, then higher up, find a zone of evergreen trees, and finally find a zone containing the snow-capped top. Cast this idea on urban air space; zone the entire indefinitely high urban air column extended vertically from the city footprint. Currently, we plan the surface, as a whole, with only limited system-wide planning of air space. Analyze the whole volume, including all air space, as a unit. Make zones accommodating buildings of varying heights; zone for pathways for drones; set speed limits for such; create a legal structure that is associated with true 3D planning, as altitudinal zonation of the urban landscape (Arlinghaus, 1986). What are the implications for mapping, archives, and timelines?
- What sorts of 'big-picture' constructs from pure mathematics of the past 50 to 100 years will find their way into the foundations of spatial thinking? A few examples come to mind: category theory, sheaf theory, and topos theory.

Kerski:

- Will the widening use of story maps and other multimedia maps lead to a greater understanding of our complex planet, and greater respect for the environment and for people?
- Will the expanding use of maps of all kinds lead to more spatial thinking and analysis throughout primary, secondary, university, and informal education? What metrics should be developed to measure this? How can we encourage researchers to tackle this important research project?
- As maps become more numerous, commonplace, and easy to create, will people become critical thinkers with regard to using and assessing maps in the future, or simply consume any map that they see without questioning it?
- How will location privacy affect the use of mapping technologies in the future?
- With the advent of enterprise and Web GIS, will spatial analysis through geotechnologies truly be as common someday as working with spatial statistics or even word processing and spreadsheets? How much should those in language arts, business, economics, engineering, mathematics, history, and other fields who are becoming interested in using geographic information systems (GIS) be required to know about the fundamentals of GIS?
- How will GIS and GIScience change in the future? Will GIS lose its 'special and spatial' characteristics as it becomes easier to use and a part of the overall set of tools that professionals in non-GIS fields use?
- With the emergence of data science programs across higher education, how can we as a society ensure that the 'geospatial' component is embedded in such programs?

(*Continued*)

- Will we emerge in the middle of a twenty-first century in which the 'where' component is routinely used in decision-making from the local to the global scale? How can we ensure that the tools that decision-makers need are embedded in core GIS technology and workflows? Will these decisions lead to a more sustainable planet and healthier, happier people?
- Aligned with Arlinghaus's question about backing up our achievements, how will digital maps be archived? How *should* they be archived so that the ones that are created this year may be examined 10 years hence? Galleries such as those on https://storymaps.arcgis.com are transitory; the content rotates with new events and new map types. How can we ensure that this content endures? Can the capabilities of the Wayback Machine (https://archive.org/web) be extended so that it is easier to search and find these digital maps, or should a new archived library be created with search tools? Who should be in charge of such a project? Who is willing to step forward and tackle this important issue?

Larimore:

- When I first conceived of a GEOMAT as a web architecture, I did so in support of my general interests in conflict resolution. I sought for a way for people from different cultures, in conflict, to be able to communicate and explain viewpoints in a non-threatening manner. The unique features of GEOMAT are using the linking feature of the internet anchored to a central timeline and base map matrix to create an indefinitely flexible array of content-filled web pages. Also, all timelines have a calendrical format, and base maps use either miles or kilometers, thus allowing accurate comparisons. Only with good communication that is interactive in nature can such resolution come about. The GEOMAT was one tool designed using recently introduced technology. Later, I was delighted to see the story map format developed with similar goals. With these tools offering progress, I look forward to seeing how future technological innovations will further enhance opportunities for conflict resolution and a more peaceful world where individuals of diverse cultures can live together equitably.

Naud:

- As cities seek to solve climate change, how does spatial thinking in environmental contexts support community understanding, investment, and behavior change?
- In the future, will we be linking person-level health care data with spatial-environmental data to demonstrate the health outcome effects of tree canopy (or even tree species) differences?
- If we can better demonstrate the health outcomes of city capital investments in parks, bike paths, sidewalks, and safe routes to school, can we better allocate costs to those that benefit, such as health insurance plans?
- Will we be able to model what the community will look like in 50–100 years if new trees are not planted and existing trees are not maintained on private property? Can we predict health outcomes under a variety of these 50- to 100-year project scenarios?
- Can we find ways to use more granular data to look for outcomes from urban planning decisions on how the community uses its spaces and how the interaction of people and the environment over time inform future capital investments?

From there, we challenge the reader to make his/her own list. Keep it. Add to it over time. Convert it to a GEOMAT or a Story Map. Check back to see how it dovetails, or does not, with reality! Track your progress with spatial thinking tools: maps, archives, and timelines.

Reference

Arlinghaus, S. L. 1986. Measuring the vertical city. In *Essays on Mathematical Geography*, pp. 39–49. Ann Arbor, MI: Institute of Mathematical Geography. http://www.imagenet.org.

Index

D

E

F

I

K

L

M

N

O

P

Q

R

S

T

U

V

W

Y

Z